Knowledge and Innovation for Competitiveness in Brazil

WBI Development Studies

Knowledge and Innovation for Competitiveness in Brazil

Alberto Rodríguez

with
Carl Dahlman and Jamil Salmi

THE WORLD BANK
Washington, DC

ISBN: 978-0-8213-7438-2
eISBN: 978-0-8213-7439-9
DOI: 10.1596/978-0-8213-7438-2

Cover design: Bill Pragluski, Critical Stages

Library of Congress Cataloging-in-Publication Data
Knowledge and innovation for competitiveness in Brazil / edited by Alberto Rodriguez, Carl Dahlman, Jamil Salmi.
    p. cm.
  Includes bibliographical references and index.
  ISBN 978-0-8213-7438-2—ISBN 978-0-8213-7439-9 (electronic)
  1. Technological innovations—Brazil. 2. Labor supply—Effect of technological innovations on—Brazil. 3. Human capital—Brazil. 4. Brazil—Economic policy. I. Rodriguez, Alberto, 1965– II. Dahlman, Carl J., 1950– III. Salmi, Jamil.

  HC190. T4K66 2008
  338'.0640981—dc22

                                                                        2008011416

Contents

Tables

Foreword

Knowledge and innovation have always been important drivers of human and economic development, and they have taken on a larger role in recent decades with the acceleration of technological change and globalized communication and trade. It is in this context that Brazil has begun to ask itself an increasingly urgent question: Why is it not growing at the rates of other middle-income countries such as China, India, and the Russian Federation? The government has pursued successful policies to halt decades of hyperinflation and pay down external debts. Private sector companies have honed their competitive edge, and some have ridden the longest commodity boom in decades to positions of global dominance. And yet, despite these remarkable achievements, Brazil remains mainly an exporter of raw materials whose economic growth has averaged only 2.5 percent per year over the last 10 years—less than half the pace of China, India and Russia.

This study provides a broad, cross-sectoral analysis of Brazil's capacity for producing knowledge and innovation. As such, it moves beyond the traditional recommendations—that is, build a stable macroeconomic environment and business-friendly physical and policy infrastructure—and instead seeks a more comprehensive approach. The fact is that Brazil has delivered some important successes with efforts to develop innovation in agriculture, aerospace and energy. But like other middle-income nations, it is discovering that it must re-evaluate its education system, its information technology infrastructure, and its policy framework for encouraging innovation to ensure that its economy as a whole is growing fast enough to keep up with the global competition while also guaranteeing progress in its fight against poverty.

The process of conceptualizing this study began with a benchmarking exercise using an analytical framework, developed by the World Bank Institute, for evaluating countries' readiness to successfully integrate into the global knowledge economy. The four pillars of that framework are (1) an educated and skilled population, (2) an effective incentive and institutional regime, (3) an efficient innovation system, and (4) a dynamic information infrastructure. Good performance in all areas is generally required for a country to take advantage of new opportunities.

In the analysis of Brazil that followed, three main issues emerged. First, Brazil trails its counterparts, particularly in Asia, in providing a quality education to all citizens. Second, Brazil has consistently aimed for the kind of elite, capital-intensive innovation that produces world-class technological breakthroughs while overlooking the kind of day-to-day innovations in production processes that tend to deliver the greatest economic returns. Third, Brazil has relied too heavily on government leadership to foster innovation, while overlooking the more cost-effective approach of using incentives to encourage private sector innovation, which tends to spread more quickly through the broader economy. In addition, trade barriers have not been eliminated fast enough to open the private sector to the kind of global competition that is necessary to foster innovation.

This study was developed in close consultation with Brazilian government and civil society leaders, who are deeply engaged with the question of how to foster innovation and greater economic competitiveness. Indeed, the breadth of the support for this study is a testament to Brazil's pragmatism and perseverance in pursuing more robust growth. It is also a welcome reflection of its continually evolving relationship with the World Bank. Today, Brazil has emerged as a leader of efforts to build South-South cooperation. In this role, it can set an important example for other middle-income nations and act as a bridge between the northern and southern hemispheres.

Developing a policy framework to foster innovation is no easy task, but experience shows that countries such as South Korea and Ireland have made great leaps forward in just a decade. The most successful innovation programs are based on a well-articulated vision, societal agreement around the program, and efforts to address all four pillars of the knowledge economy through a combination of bottom-up initiatives and top-down reforms. We hope that this volume will contribute to Brazil's efforts to analyze its strengths and weaknesses and chart an effective way forward, for itself and for other middle-income nations. In taking on this challenge, Brazil leads the way for many countries that might also benefit from a close examination of their capacity to innovate and compete.

Rakesh Nangia   John Briscoe
Acting Vice President   Country Director for Brazil
World Bank Institute   The World Bank

Acknowledgments

This study is the result of the collaborative efforts of a team of World Bank staff and consultants inside and outside Brazil. Alberto Rodríguez, Lead Education Specialist in the Human Development Department for the Latin America and the Caribbean Region (LAC), was the lead author and manager of the work. Co-authors Carl Dahlman of Georgetown University and Jamil Salmi of the World Bank's Human Development Network provided major contributions without which this study would not have been possible.

The authors wish to thank the many World Bank staff members who contributed their expertise and dedication to this project. Core team members included José Guilherme Reis, Senior Private Development Specialist; Anuja Utz, Senior Operations Officer, World Bank Institute; Julio Revilla, Senior Economist; and Junior Professional Associates Mariam Dayoub and Domenec Devesa.

Special thanks go to the World Bank Institute's Knowledge for Development (K4D) Program, which provided the benchmarking analysis of Brazil's transition to the knowledge economy. This analysis formed the basis for early consultations and debate on education and competitiveness with key Brazilian stakeholders. The K4D team also provided input for several chapters of the book.

The authors are grateful to have received the generous support and guidance of John Briscoe, the World Bank's Country Director in Brazil. Significant contributions were also made by World Bank staff members Andreas Blom, Education Economist; Paulo Correa, Senior Economist; José Luis Guasch, Senior Advisor; Keiko Inoue, Operations Officer; Jenny Litvack, Lead Economist for Human Development in LAC; Eduardo Vélez, Education Sector Manager for LAC; Ethan Weisman, Lead Economist for Brazil; Zeze Weiss, Senior Civil Society Specialist; and Alexey Volynets, Operations Analyst, International Finance Corporation; and by consultants Jefferey Marshall, Horacio Hastenreiter, Janssen Teixeira, Chloe Fevre, Jason Hobbs, Andrea Welsh, and Antonio Magalhães.

Several reviewers offered invaluable feedback and advice throughout the conceptualization and writing stages. These included Kathy Lindert, Christopher Thomas, Alfred Watkins, Robin Horn, and William Maloney. Sheldon Annis contributed his editorial expertise to the project, and the World Bank Office of the Publisher produced the English-language book. Brazil's National Confederation of Industry, the CNI, translated the volume and produced the Portuguese-language version for distribution in Brazil.

Finally, the authors wish to thank all the Brazilian authorities at the federal and state levels who supported this study, as well as the many private sector, civil society, and education sector representatives who contributed ideas, information and feedback throughout the research and writing process.

Abbreviations

ADBI	Brazilian Agency of Industrial Development
ANPEI	Associação Nacional de P,D&E das Empresas Inovadoras (National Association of Research, Development, and Engineering in Innovative Firms)
BNDES	National Bank for Economic and Social Development
BRIC	Brazil, Russia, India, and China
BRICKM	Brazil, Russia, India, China, Korea (Rep. of), and Mexico
CAPES	Coordenação de Aperfeiçoamento de Pessoal de Nível Superior (Coordination for the Improvement of Higher Education Staff)
CBPF	Centro Brasileiro de Pesquisas Físicas (Brazilian Center for Research on Physics)
CCT	National Council on Science and Technology
CCTs	conditional cash transfers
CEB	Censo do Capital Estrangeiro no Brasil (Central Bank's Foreign Capital Census)
CEF	Caixa Economica Federal (Federal Savings Bank)
CEPEC	Center for Research in Education Culture and Community Action
CGEE	Center for Management and Strategic Studies
CNA	Confederation of Agriculture and Livestock
CNC	National Confederation of Business
CNDI	National Council of Industrial Development
CNI	National Confederation of Industry
CNPq	Conselho Nacional de Desenvolvimento Científico e Tecnológico (National Council for Scientific and Technological Development)
EFA	Education for All
EJA	Educação de Jovens e Adultos (Education for Youth and Adults)

Embrapa	Empresa Brasileira de Pesquisa Agropecuária (Brazilian Agricultural Research Corporation)
ENADE	Exame Nacional de Desempenho de Estudantes
ENCEJA	Exame Nacional de Certificação de Jovens e Adultos
ENEM	Exame Nacional do Ensino Médio
EU	European Union
FAPESP	Fundação de Amparo à Pesquisa do Estado de São Paulo (São Paulo State Research Foundation)
FDI	foreign direct investment
FIES	Fundo de Financiamento ao Estudante do Ensino Superior
FIESP	Federation of Industries of the State of São Paulo
FINEP	Financiadora de Estudos e Projetos (Financier of Studies and Projects)
FNDCT	Fundo Nacional de Desenvolvimento Científico e Tecnológico (National Fund for Scientific and Technological Development)
FPR	Rural Professional Training
FUNDEB	Fund for the Development of Basic Education
FUNDEF	Fund for the Development of Fundamental Education & Valorization of Teachers, also known as FVM
GCI	Global Competitiveness Index
GDP	gross domestic product
GNI	gross national income
IBGE	Brazilian Institute of Geography and Statistics
IC	investment climate
ICS	Investment Climate Survey
ICT	information and communications technology
IDB	Inter-American Development Bank
IIT	Indian Institutes of Technology
IMPA	Instituto Nacional de Matemática Pura e Aplicada (National Institute of Basic and Applied Mathematics)
INEP	National Institute for Education Research and Study
Inmetro	Instituto Nacional de Metrologia (National Institute for Metrology)
INPI	Instituto Nacional de Propriedade Intelectual (National Institute of Intellectual Property)
IPEA	Institute of Applied Economic Research
IPI	Imposto sobre Produtos Industrializados
IPR	intellectual property rights
IPT	Instituto de Pesquisas Tecnológicas do Estado de São Paulo (São Paulo State Institute for Technological Research)
ISO	International Standards Organization
K4D	Knowledge for Development
KAM	Knowledge Assessment Methodology (World Bank)
LAC	Latin America and the Caribbean region
LDB	Lei de Diretrizes Básicas (National Education Law)

MCT	Ministry of Science and Technology
MDIC	Ministry of Development, Industry, and Trade
MEC	Brazilian Ministry of Education and Sports
MIT	Massachusetts Institute of Technology
MLE	medium and large enterprise
MSE	micro- and small enterprise
MSTQ	metrology, standards, testing, and quality control
NGO	nongovernmental organization
NRI	Network Readiness Index
OECD	Organisation for Economic Co-operation and Development
PACTI	Programa de Apoio à Capacitação Tecnológica da Indústria (Support Program for Technological Industrial Training)
PADCT	Programa de Apoio ao Desenvolvimento Científico e Tecnológico (Support Program for Scientific and Technological Development)
PAM	Mobile Activities Program
PBQP	Programa Brasileiro da Qualidade e Produtividade (National Program for Quality and Productivity)
PDDE	Projeto Dinheiro Direito na Escola
PDE	Plano de Desenvolvimento da Educação (Plan for Educational Development
PDTI/PDTA	Programas de Desenvolvimento Tecnológico Industrial e Agropecuário (Programs for the Development of Industrial and Agricultural Technology)
PIA	Pesquisa Industrial Anual (IBGE's Annual Industrial Survey)
PINTEC	Pesquisa Industrial–Inovação Tecnológica (IBGE's Industrial Survey–Technological Innovation)
PISA	OECD Programme for International Student Assessment
PME	Monthly Employment Survey
PNAD	Pesquisa Nacional por Amostra de Domicílios (National Household Survey)
PNC	National Curriculum Parameters
PPP	purchasing power parity
PPV	Standard-of-Living Survey
PROEP	Program for the Reform and Enhancement of Professional Education
PROME	Program for the Enhancement and Expansion of Secondary Education
ProUni	Programa Universidade para Todos
PS	Social Promotion
PVA	potential value added
R&D	research and development
RAIS	Relação Anual de Informações Sociais (Ministry of Labor and Employment's Annual Listing of Social Information)
RCA	revealed comparative advantage
S&T	science and technology

SAEB	National System for Basic Education Evaluation
SE	small enterprise
SEBRAE	Serviço Brasileiro de Apoio às Micro e Pequenas Empresas (Brazilian Service for Assistance to Small Business)
Secex	Secretaria de Comércio Exterior (International Trade Secretariat)
SENAC	National Service for Commercial Apprenticeship
SENAI	Serviço Nacional de Aprendizagem Industrial (National Service for Industrial Apprenticeship)
SENAR	National Service for Agriculture Apprenticeship
SENAT	National Transport Apprenticeship Service
SES	socioeconomic status
SESC	Social Service for Commerce
SESCOOP	National Apprenticeship Service in Cooperative Activities
SESI	Social Service for Industry
SEST	Social Service for Transport Industries
SESU	Ministry of Education's Secretary for Higher Education
SINAES	Sistema Nacional de Avaliação da Educação Superior
SMEs	small and medium enterprises
STI	scientific and technological institutions
TFP	total factor productivity
THES	*Times Higher Education Supplement* (United Kingdom)
TIMSS	Trends in International Mathematics and Science Study
TVET	technical and vocational education and training
VET	vocational education and training
UNDIM	National Association of Municipal Education
UNDP	United Nations Development Programme
UNESCO	United Nations Educational, Scientific, and Cultural Organization
UNESP	Universidade Estadual Paulista Júlio de Mesquita Filho
UNICAMP	Universidade Estadual de Campinas
USP	Universidade de São Paulo
WDI	World Development Indicators
WEF	World Economic Forum
WIPO	World Intellectual Property Organization

Currency Equivalents
(Exchange Rate Effective: June 19, 2007)

Currency Unit = Real
R$1.00 = US$0.53

Executive Summary

Brazil has made considerable progress toward macroeconomic stability since reform measures began to take hold in the early 1990s, and its economy has produced stronger growth as a result—an average of 2.5 percent annually over the past decade. Nevertheless, from an international perspective, Brazil's level of economic growth is still a matter of significant concern. Compared with Organisation for Economic Co-operation and Development (OECD) countries or with competitors such as China or India, Brazil not only is growing slowly, it is falling farther behind. Indeed, as shown in figure ES.1, the income gap between Brazil and OECD countries has substantially widened. In 1980, Brazil's per capita purchasing power parity was about 42 percent of that of OECD countries. Twenty-five years later, it had fallen to under 29 percent of OECD countries.

Where Growth Comes From

Economic growth is widely understood as the interaction between physical and human capital. Investment in either generally increases growth; moreover, when physical and human capital interact more efficiently, growth occurs more rapidly. Economists generally attribute this incremental efficiency-based growth to total factor productivity (TFP). During the exceptional high-growth era of the "Brazilian Miracle" (1960–80), TFP was critical to growth; however, since then, TFP has declined dramatically. Growth-accounting exercises show that the ratio of Brazil's TFP compared with that of the United States dropped from 1.07 in 1975 to 1.02 in 1980, to 0.80 in 1995, and to 0.73 in 2000.

The macroeconomic shocks of the 1970s and the debt crisis of the 1980s are important factors in explaining the slowdown in Brazil's growth. However, this report argues that the decline in TFP was a similarly important cause. Why did it happen? Brazil's low rate of investment is one part of the answer.

Figure ES.1. Brazil's Per Capita Income Relative to the OECD Area (in PPP)

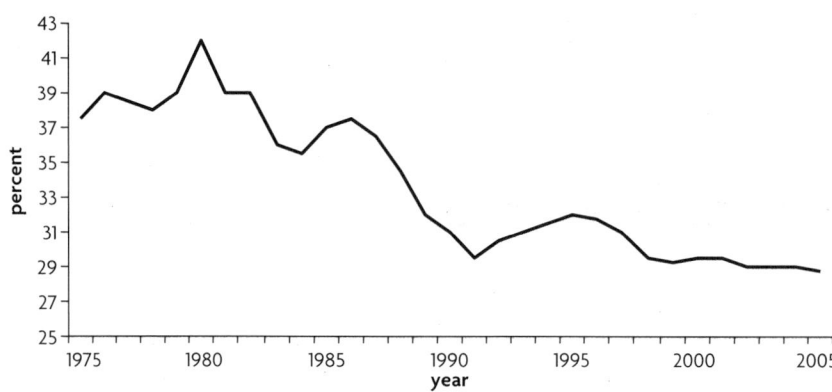

Source: Based on data from the OECD Web site (http://www.oecd.org).

Low productivity is another. The main factor, however, is that a new global "knowledge economy" has been emerging; and Brazil, despite its relatively successful implementation of adjustment policies in the mid-1990s, was not prepared to compete.

In the new paradigm for middle-income countries, knowledge—not natural resources or cheap labor—increasingly constitutes the core of a country's comparative advantage. As well illustrated by dramatic success stories such as that of Bangalore, the capital of the Indian software industry, technical innovation and knowledge can work hand-in-hand to lead a country from suffocating poverty to strong productivity and competitiveness. Indeed, the proportion of goods in international trade with a medium-high or high technology content rose from 33 percent in 1976, to 54 percent in 1996, and to 64 percent in 2003 (World Bank 1999). This period was the same one during which Brazil muddled through slow trade liberalization and weak labor reforms and paid little attention to its lagging basic education system. Had more radical reforms been undertaken, Brazil would have been much better able to take advantage of domestic and international opportunities to spur growth, as did competitors such as China.

Brazil can no longer ignore the knowledge economy—and it is not. An ongoing national dialogue is taking place on reforms to sustain strong macroeconomic performance, further open trade, improve the physical infrastructure, strengthen the judicial system and legal environment, and deal with weak and inequitable education systems that are not producing the kind of human capital required by today's global competition. This report emphasizes that Brazil has indeed made significant progress; yet the hard reality is that Brazil's competitors have too—only faster. The question has become not only how Brazil can make further progress but also how it can catch up.

The analysis in this report is based on the conceptual framework shown schematically in figure ES.2. Following from the conceptual framework, the report discusses three main areas for enhancing competitiveness and

Figure ES.2. A Conceptual Model for the Components of Growth

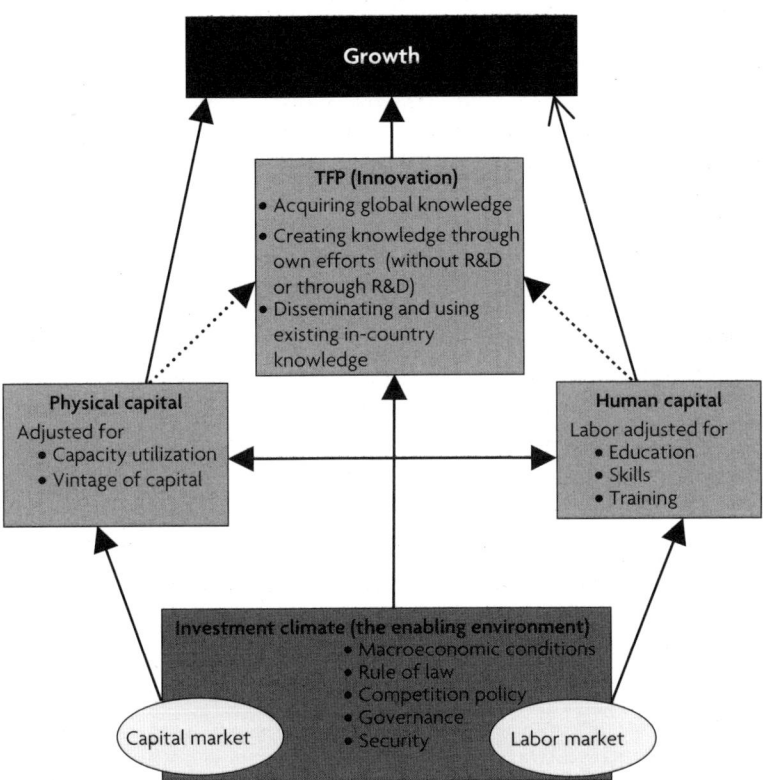

Source: Authors.

accelerating growth. First, Brazil needs to build upon its stable macroeconomic environment to extend reforms that will improve the investment climate. Second, higher productivity will require a focused effort to expand TFP through innovation-based growth. Third, a series of "micro" reforms are needed, of which two are urgent—strengthening incentives for firms to innovate, and upgrading the education system to improve the skills of workers entering the labor force. We present a set of specific recommendations that stem from this analysis. We also discuss possible roles for Brazilian agencies in implementing these recommendations, as well as the need to raise awareness on the urgency of this agenda.

The main messages of this report cover four topics—consolidating the macroeconomic environment, boosting innovation, improving skills across the labor force, and moving from analysis to action.

Consolidating the Macroeconomic Environment

The report summarizes key conclusions from previous World Bank policy papers on the macroeconomic fundamentals behind Brazil's current stability and progress. It discusses improvements in the enabling environment that would serve to drive accelerated growth.

The Brazilian economy has remained stable as a result of prudent macro-economic management—including fiscal and monetary policy, as well as debt management. Improved macroeconomic fundamentals have reinforced the benefit of favorable external demand for Brazil's primary commodities, raising international reserves to unprecedented levels. Fiscal restraint, which has included a cap on public investment, has translated into yearly primary surpluses and macroeconomic stability. However, the country's infrastructure now needs upgrading in order to increase productivity and avoid jeopardizing growth.

The challenge facing Brazil is to continue reducing public debt and improving the quality of the fiscal adjustment (that is, ensuring adequate resources for key public investments and poverty alleviation programs) while also improving the efficiency of public expenditures to create the fiscal space necessary for pro-growth investments. The ability of the government to adjust the composition of public expenditures is constrained, however, by its current high level of spending (most notably on pensions) and by an ongoing debt burden that ultimately limits the government's borrowing and spending capacity. In addition, the continuous growth in the size of government during the past decade—financed through increased taxes—has constrained domestic savings. High interest rates have acted as a disincentive to private sector investment.

In short, a stable macroeconomic environment has helped to reverse the bitter declines of the so-called "lost decade" of crisis and stagnation in the 1980s, and this has led to moderate growth in the past few years. However, a stable macro environment has not been sufficient to spark *fast* growth. Moreover, given inadequate public investment in infrastructure and the sluggishness of reforms to facilitate the investment climate, prospects for significantly higher growth remain slim. While productivity improved during the past decade, as shown by historical evidence in this report, it is nonetheless lower than in previous periods when investment grew faster.

Boosting Innovation

Brazil's growth depends strongly on the export of manufactures and commodities, a dependence that is likely to continue. Yet with few exceptions, Brazil's manufacturing base lags with respect to innovation—especially when Brazil is compared with China or India, countries that have taken giant steps in growth-enhancing innovation. If recent trends continue, Brazil would continue to be mainly a supplier of primary commodities in world markets and an exporter of manufactured products to Mercosur and other Latin American countries. In other words, Brazil risks missing the opportunity to become a serious, diversified, global competitor, which would require it to emphasize higher value added in products in the sectors in which it already has some comparative advantages, and to engage in higher-value, more-income-elastic manufactures and services. Brazil needs not only to diversify and add value to its commodities, but also to improve its competitiveness in manufacturing and service exports as well.

Until the 1990s, the productive sectors in Brazil operated within a relatively protected economy. The government provided few incentives for private sector investment in innovation; yet that mattered less because protection from competition made private sector investment in innovation relatively less necessary. We argue in this report that two factors—a bias toward overly "theoretical" research in publicly funded universities and significant underinvestment by a shielded private sector that is spared the need to compete—lie at the heart of Brazil's current relative underperformance in innovation.

The private sector needs to invest more in R&D. Recent initiatives to encourage firms to invest in innovation—for example, the Innovation Law and the Sector Funds—are welcome steps. However, as argued throughout this report, the government now needs to take these measures further by creating a broader enabling environment in which private firms are willing to invest in innovation, take risks, and expand their productive activities into new, "less-safe" areas. In addition to increasing its overall investment rate, Brazil needs to further liberalize the economy, in part, to force firms to become more competitive.

Public investment in R&D needs to be made more effective, not just by producing more knowledge and technology but also by providing the infrastructure to commercialize and disseminate new knowledge (for example, technology parks, technology transfer offices, business incubators, and venture capital operations). Spain provides a notable example of how such efforts can work. Moreover, as we argue below, Brazil also must invest more in human capital through quality basic education and advanced skills training. China, Ireland, the Republic of Korea, and Singapore are just a few of many examples where this has been done massively and successfully.

This report proposes a broad new definition of innovation. As used here, the term refers not just to new products and processes but also to new business processes and new ways of carrying out productive activities. We emphasize that innovation to improve TFP should not be understood simply as invention or the first use globally of a new technology but also as the first application of a product or process in a specific setting. Because developing countries are behind the technological curve in most sectors, they need to think less about invention and more about doing things differently with available knowledge and technology that they can acquire. The report proposes a three-stranded typology of innovation: (a) creation and commercialization of new knowledge and technology, (b) acquisition of knowledge and technology from abroad for local use and adaptation, and (c) the dissemination and effective application of knowledge and technology (whether domestically created or acquired from abroad) that is already available in-country though not broadly utilized. The significance of these distinctions is discussed below.

Creating and Commercializing New Knowledge and Technology. In Brazil, investment in technological innovation comes mainly from the public sector—about 55 percent of the total, compared with about 30 percent in the United States. A research culture that is heavily and reliably financed by the public sector has excelled in the production of conceptual knowledge—for example, Brazil accounts for nearly 2 percent of articles published in internationally recognized research journals (roughly on par with Brazil's 2 percent of world gross domestic product [GDP]). On the other hand, substantial public expenditure has been far less successful at energizing technological innovation—for example, patents that can be commercialized. According to the World Intellectual Property Organization (WIPO), Brazil accounted for about 0.18 percent of patents in 2000. This compares with 3.4 percent of patents attributable to Sweden—that is, nearly 19 times more patents than Brazil despite a much smaller population. Similarly, Korea accounted for 1.7 percent of patents, more than nine times the rate for Brazil.

Ironically, Brazil invested in R&D infrastructure far earlier than most other developing countries. Yet this report finds that an intellectual and practical disconnect has now emerged in Brazil that is not always found elsewhere. The public universities and labs where most government-funded research is conducted primarily pursue "pure" conceptual knowledge. Links between the private sector and these universities and labs are not well developed, unlike in other countries where entrepreneurial scientists and engineers typically have a foot in both worlds. Moreover, the private sector's own research capacity has been diminished by underinvestment from companies protected by trade barriers from foreign competition. The net result is that Brazil needs to pay far greater attention to what is produced through public investment, what happens to new knowledge once it is created, and how the private sector can be mobilized as an active partner. Strengthening the institutions and norms that protect intellectual property and supporting business incubators would help immediately.

A nation's capacity to create new knowledge and technology is closely associated with advanced technical skills and a tertiary education system that is particularly strong in science, engineering, and technology application. Brazil has emphasized the humanities and social sciences at the expense of science and engineering. Despite slow but steady growth in the latter disciplines, Brazil's tertiary education system still has far too little capacity to train advanced innovators who can work at the frontier of global knowledge creation. In China, the government has tapped and supported both public and private universities to increase enrollment rapidly and to leverage respective comparative advantage. As Brazil wrestles with the coverage, relevance, and resource needs of its higher education system, the Chinese examples could be instructive.

Acquiring and Adapting Global Knowledge and Technology. For countries not already on the cutting edge, it is generally more practical to acquire rather than invent new knowledge and technology. Transfer of technology can be accomplished through several means—direct foreign investment; licensing; technical assistance; technology embodied in capital goods, components, or products; copying and reverse engineering; foreign study; published technical information, especially on the Internet; twinning; cooperative training partnerships; distance learning; and more. Trade—specifically, importing the latest versions of hardware, machinery, and software—is probably the most direct and critical means of acquiring knowledge and technology. Brazil is still struggling to reconcile the relative comforts of protectionism with the inevitable need to compete in global markets. In this respect, Brazilian firms are just awakening to the full benefits that acquired foreign technology can bring. Not surprisingly, the firm-level analysis of innovation undertaken for this report found large firms (and especially multinational firms) to be far ahead in innovation and productivity.

The capacity of firms to put acquired technologies to productive use points again to the challenges of human capital formation. Technology stands little chance of being adopted and adapted successfully if workers lack basic reading and math skills; or at a higher level, the ability to reason conceptually, think outside the box, and apply the scientific method. Workers with these skills are no less critical than higher-level managers who can quickly adjust to computerization or imaginatively redesign a production strategy. If firms cannot trust in the adaptability of their employees, they necessarily become risk averse, opting for the low road to economic survival—that is, heavier exploitation of cheap, unqualified labor (as we found occurring in the northeast of Brazil). In essence, both basic and advanced skills are needed for a firm to maximize the rewards of acquired innovation.

Disseminating and Using Knowledge and Technology That Is Already Available In-Country. Firms' inputs, processes, and outputs were disaggregated, broken down by sector, size, and region. Data from the World Bank Investment Climate Survey (ICS) and the Brazilian National Innovation Survey of the Brazilian Industrial Sector (PINTEC) were used for this analysis, and the results are presented in this report. Microanalysis allowed a closer look at the characteristics of firms within and between sectors, as well as comparisons with firms in other countries. Some Brazilian firms were clearly found to be innovators, mainly large enterprises with many employees and strong outputs. In general, however, Brazilian firms were found to innovate less than those of other countries. There is relatively little demand for innovation in the unsophisticated internal market. Protection continues to undercut the need for innovation and creative risk taking. Firm productivity is low, and dispersion of productivity is enormous. In fact, the report found that the dispersion in firm productivity in Brazil was much greater than in most other countries for which data were available, including China and India.

This report argues that using the knowledge already in Brazil provides the quickest and most promising route for increasing productivity and competitiveness to spur growth. Through this third type of innovation—which is arguably the least expensive and most accessible—Brazil could increase productivity across all sectors. The report underscores the critical importance of firms being able to identify productive practices *within* the country and then having the inputs to replicate, enhance, and increase their own productivity. This third kind of innovation requires relatively greater effort to disseminate knowledge through channels such as industrial and service extension programs, technical information centers, and cluster-based technology improvement programs. Some innovation requires newer machinery and better physical inputs, as well as better management and organization. Equipment is not a magic bullet, however. What matters is what happens on the shop floor. Can workers observe new practices first-hand, and is there an environment that rewards increased efficiency and productivity? Indeed, can workers accomplish the same things through better use of the equipment and inputs that they already have?

The fact that job tenure in Brazil is generally low—and lower still for less-skilled workers—might be expected to increase the flow of good practices between firms. In reality, however, this does not appear to be happening. We suggest that the lack of basic skills among workers is probably the single most significant obstacle to the use of new technology and equipment or the free flow of innovative practices across firms. Indeed, unskilled workers are likely to be risk averse and more comfortable with the simple routine of procedures that do not demand additional formal training. Moreover, high job turnover may discourage effective firm-level training. Our study found that Brazilian firms *do* invest significant time and resources training their employees; however, in most cases this training focuses on basic skills deficits that should have been addressed by the formal education system, not on the introduction of innovation to improve productivity on the shop floor.

One notable exception is the production chains that have been developed by small and medium enterprises that act as suppliers to large innovative firms such as Embraer, Petrobrás, Gerdau, Ford, and others. These smaller firms frequently are able to enhance their productivity by using technologies adapted from the larger innovative companies. Cases such as these tend to occur in specific geographic clusters. The local qualifications of human resources—both advanced and basic—are crucial to these processes, as the experience of Embraer demonstrates.

Improving Skills Across the Labor Force

Brazil's unemployment rates worsened for all workers during the 1990s—ranging from those with no education through those with primary, secondary, and tertiary education. The proportion of unemployed university graduates rose to 16.4 percent, compared with an unemployment rate of 9.3 percent for the population at large. This is highly suggestive of a mismatch between the skills of formal education system graduates and the needs of the labor market, rather than a sign that the labor market does not require advanced skills. The extremely high rate of secondary school dropout similarly reflects weakness in the school-to-work transition. Older secondary students, in particular, drop out because they know that staying in school will not necessarily provide additional opportunities for jobs or for meaningful job-oriented training. In addition, there are insufficient graduates from nonuniversity institutions and short-duration professional programs, such as those typically offered by community colleges in the United States and postsecondary technical institutes in Europe.

Strengthening Tertiary Education. It is well accepted that more and better education improves employability and earnings. However, average educational attainment for the Brazilian population age 15 and older is still only 4.3 years. With only a quarter of the university-age population attending a tertiary institution, Brazil has the next-to-lowest gross enrollment rate among the larger Latin American countries, well below the continental average of 30.3 percent. The low enrollment rate in universities is mirrored by the very small proportion of the labor force with tertiary-level educational qualifications: 8 percent.

Despite many top-quality enclaves at the tertiary level, the overall lack of consistent high quality (especially in the absence of performance standards) is critical. Brazil is the world's eighth-most-populous country, yet no Brazilian university is to be found among the 100 top-ranked universities worldwide. Research production is concentrated in a very small group of elite public or state universities. A second tier of public and private universities has many pockets of excellence, but beyond that point on the spectrum—that is, in the vast majority of small underfunded private universities—quality is worse than uneven and serious research is neither financed nor rewarded. At the federal universities, 83 percent of instructors are full-time academics, in contrast to about a third of instructors in the municipal universities and a fifth in the private institutions. In private universities, most instructors are part-time employees. Basically, they earn an hourly wage and they are paid according to the number of classes that they teach.

The proportion of academics with a doctoral degree rose from 15 percent in 1994 to 21 percent in 2004. At the federal universities, the rate doubled from about 21 percent to 42 percent. The vast majority of academics not only have not been trained in research through doctoral training, they have virtually no opportunity to participate in publicly funded basic R&D. That does not mean, however, that they are more likely to engage in "practical" research or that they engage in outside-the-university research with private sector counterparts. To the contrary, the university and private sector realms remain consistently separate across the board. Unlike the Silicon Valley or Route 128 models in the United States—where well-trained innovators may constantly shift from university to private sector and back throughout their careers, or may simply maintain a permanent presence in both—their Brazilian counterparts remain remarkably segregated. To an astonishing extent, the two worlds do not intersect, much less cross-fertilize. Similarly, only a relatively small minority of Brazilian faculty study abroad. In 2005, only 2,075 students were officially sponsored for graduate studies outside Brazil. Only 1,246 foreign students attended Brazilian universities.

Other postsecondary training is offered by private providers and, in particular, by the institutions that form the "S-system." These nine institutions constitute the largest consolidated professional training system in Latin

America, created by the National Confederation of Industry (CNI) and the state federations of industry. The system is financed through a compulsory 2.5 percent payroll tax. Present in about 60 percent of Brazilian municipalities, the S-system offers an estimated 2,300 courses per year and enrolls about 15.4 million trainees annually. While the effectiveness of its training (and the cost-efficiency of the system itself) has been hard to assess, the S-system plays a crucial role in providing specific training for workers and could serve as the cornerstone for a lifelong learning framework in Brazil.

Access to tertiary education—especially at the most prestigious universities— is skewed heavily toward upper-income families. While approximately 69 percent of the population is classified as low income in Brazil, about 90 percent of students at UNICAMP (generally regarded as one of the top two universities) are *not* low income. This unequal distribution at UNICAMP is hardly unique; it reflects a continuing pattern of unequal opportunity across the system more broadly. At the secondary level, for example, about 90 percent of children from the highest income decile complete school, compared with only about 4 percent of children from the lowest decile of families.

Improving Basic Education. If a weak and relatively small tertiary education system presents a challenge for Brazil's innovation system, basic education is also at the heart of the country's low productivity and lack of competitiveness. Besides too few educational opportunities in the absolute sense (and setting aside the social inequities of who benefits), the Brazilian education system is significantly deficient in the quality of education that it offers. As shown in this report, schools at the primary and secondary levels are failing to provide the minimum literacy and numeracy skills necessary for active citizenship, let alone productive participation in a technology-based labor market. According to the international PISA tests, approximately half of Brazilian 15-year-olds have difficulty reading or cannot read at all, and about three-fourths cannot manage basic mathematical operations. It is therefore unsurprising that this report found that, while Brazilian firms invest significant resources in worker training, these efforts are mostly geared toward filling the basic skill gaps left by the formal education system. Companies should be building upon basic skills, not having to provide them.

As discussed in the report, there are many reasons for the unsatisfactory performance of the nation's schools, including the management and incentives of the teaching profession. Relatively speaking, Brazil's 1.5 million teachers are reasonably well paid. They earn 56 percent more than the average national salary overall. (By contrast, teachers in OECD countries on average earn about 15 percent *less* than the average salary in their respective countries.) The pay gradient for Brazilian teachers is tightly defined by seniority. With few exceptions, neither penalties nor rewards are available as incentives for teacher performance, much less student learning. Unsurprisingly, given the pace of enrollment expansion in recent years, funding for math, science, and technology enrichment has lagged far behind school construction and teacher hiring as a budget priority. Nearly a third of those who teach Brazil's

45 million students have not completed university training, and only about 20 percent hold master's degrees. For the most part, the training of those who are university-educated tends to be very strong in pedagogical theory but very weak in the applied art of teaching.

Over the past 20 years, the number of places in primary and secondary schools has increased dramatically, and access to primary education is now virtually universal. It is less certain, however, that the *quality* of education has increased. This is related less to absolute lack of financial resources (public educational expenditure rose from 3.9 percent of GDP in 1995 to 4.3 percent of GDP in 2005) than to management factors. For example, it is estimated that about 60 percent of school principals obtained their jobs based on political criteria. Computers in the schools (approximately 2 per 100 students compared with 28 per 100 in Korea) tend to be used by teachers and administrators, not by students, which is all the more significant for future technological innovation in a country where the vast majority of families do not have a personal computer at home.

The report also discusses the pedagogical and curricular factors that contribute to low quality in basic education. Classroom teaching at the primary level (especially in rural areas) is still conducted very much as it was a generation ago. That means students passively copy what the teacher writes on the board and are expected to learn by rote memorization, an approach that is the diametrical opposite of the kind of active learning that rewards flexible thinking, conceptual reasoning, and problem-solving skills—in other words, the very traits that adult workers need for competitiveness in a knowledge economy.

In summary, the low level and skewed distribution of education among Brazilians explains more than the oft-studied cycle of poverty and inequality. Here, we argue that basic and advanced skills are critical inputs for the nation to harness innovation, increase productivity, enhance competitiveness, and accelerate economic growth—and that these needs presently are not being met.

From Analysis to Action: Who Needs to Do What?

The report proposes concrete actions in six key areas—the enabling environment, knowledge creation and commercialization, acquisition of foreign knowledge, leveraging and dissemination of technology use, basic education and skills, and tertiary education (advanced skills). Taken together, these recommendations represent a first step toward a comprehensive national plan for innovation. Continued analysis, increased public awareness, and a vigorous national debate can translate these recommendations into an integrated national strategy to foster innovation-led growth.

Leveraging innovation for economic growth necessarily encompasses a broad spectrum of issues and actors. This ranges from the overarching framework of the economic and institutional regime to highly technical, specialized applications relating to R&D, foreign investment, and technology transfer; information technology; standards and quality control; finance and venture capital; education; and so forth. The final chapter recasts the broad array of recommendations from the perspective of which actors need to take what actions. The chapter addresses the many entities of government, the private sector, and civil society that will have to implement recommendations if ideas are to be translated first into action and then into reality.

Not all of the recommendations are of equal weight and priority; and for technical or political reasons, some will be far more difficult to implement than others. Some actions would require new laws through Congress. Some would require significant changes in policies or the regulatory environment, while others could be achieved by exerting a reasonable amount of political will. Some could be carried out with existing resources. Others would require significant mobilization of public and private funds. Some actions could be done rapidly. Others will require years of sustained efforts. Some actions will be difficult because they affect the interests of groups who benefit from the system the way it is.

Our work does not go so far as to prioritize or suggest details for a particular plan. That is necessary—including all the hard choices and tradeoffs that concrete action implies—though it is beyond the scope of the present report. What is clear is that Brazil needs to undertake a broad, systemic reform process in order to increase the competitiveness of its economy and to accelerate growth. There is a danger that the recently improved trade performance—driven by the current boom cycle in commodity prices—will improve economic performance enough to temporarily justify complacency. Given the fundamental changes that are taking place globally, that short-sighted approach would be costly.

Neither the government nor Brazilian society as a whole appears to be fully cognizant of the international trends or the opportunity costs of failure to respond. Most governments and citizens of Asia *do* understand these trends, and they *are* responding, and that is an important reason why Asia is rising as the new base of economic power. For Brazil, the next step is to mobilize a mass campaign to raise public awareness. Brazil needs to see its performance in the broader global context, to analyze the new global challenges that it faces, and to discuss in a transparent way what must be done. The process of stocktaking and building stakeholder awareness is inherently a domestic political process. It needs to be locally driven and locally owned. It is hoped that this report will provide useful input into launching such a process.

Methodology and Organization of the Report

This report was carried out by a multidisciplinary team of World Bank staff, consultants, and Brazilian counterparts. The core team and contributors analyzed existing data, developed conceptual and econometric models, and consulted extensively with federal and subnational governments, business leaders, and academics. The research was conducted primarily between November 2006 and April 2007.

The authors relied on secondary data analyses by Brazilian researchers, international colleagues working in other countries on similar topics, and work by the core team itself. For the growth analysis and decomposition, the main source of data was Brazil's Geography and Statistics Institute, the IBGE, including the modified growth calculations from March 2007. For national-level analyses on innovation, the team used readily available information and databases from the Ministry of Science and Technology, the World Bank (such as the *World Development Indicators* and the World Bank's KAM interactive database), the Ministry of Finance (such as SIAFI, the Integrated Financial Administration System), and other agencies (including the U.S. Patent and Trademark Office). For firm-level analyses, the authors relied on the World Bank Investment Climate Survey, the IBGE's PINTEC Technological Innovation Survey, and a data set developed by the Institute for Applied Economic Research, IPEA, which combines firm-level data with workforce data from the Ministry of Labor. For analyses of human capital, sources included student assessment data sets and the National School Census from the Institute for Educational Research (INEP), a unit of the Ministry of Education; the IBGE's PIA Annual Industry Survey database; OECD's PISA database; and the RAIS (Annual Social Information), a database managed by the Ministry of Labor.

In some cases, the authors performed original econometric work to assess relationships and confirm the conceptual framework. In others, the study reports on econometric work designed and performed elsewhere. Recent research by IPEA, which uses a newly assembled database combining firms' and workers' information, was found to be particularly useful.

The findings of this report are organized into eight chapters, followed by several appendixes. The first chapter looks at the central problem—why Brazil has grown so slowly despite relative success in improving its fiscal and macroeconomic performance. The second chapter presents the four-factor conceptual model used to analyze economic growth, highlighting the importance of innovation and TFP. Each element of the conceptual model is analyzed separately in subsequent chapters. The third chapter defines the concept of innovation as elaborated in the study. Three kinds of innovation are distinguished—first, creation of new knowledge and technology; second, acquisition of new knowledge and technology (often from elsewhere); and third, wholesale adoption, adaptation, and dissemination of new knowledge and technology within the national economy. Applying these distinctions, the fourth chapter assesses Brazil's performance in innovation at the national level. The fifth chapter provides a similar kind of analysis at the micro level of the firm. The chapter elaborates on the relationships among innovation, productivity, and growth—and more specifically, it points to evident weakness in human capital formation. The sixth chapter looks more closely at the multi-tiered education systems primarily responsible for human capital formation. Although Brazil has a very large, nominally literate population, its workforce at every level is nevertheless poorly prepared for innovation. The chapter explains this through summary profiles of the primary education system, the secondary education system, out-of-school advanced training, and the tertiary education system. It also explores features related to school performance and governance and issues related to teachers and teaching. The primary, secondary, tertiary, and out-of-school systems are described in greater detail in accompanying appendixes at the end of the report. The seventh chapter looks broadly at what Brazil can do to foster innovation. With an eye toward developing an integrated national strategy, it proposes concrete actions in six key areas—the enabling environment, creation and commercialization of knowledge, acquisition of foreign knowledge, leveraging and dissemination of technology use, basic education and skills, and tertiary education. The final chapter reframes these recommendations from the pragmatic viewpoint of who needs to do what.

CHAPTER 1

Brazil's Growth and Performance in a Global Context

Brazil has achieved relative economic stability and a growth rate of about 2.5 percent over the past decade. However, Brazil has not recovered the rapid growth rates it once achieved, nor the current rapid growth rates of its main global competitors. In fact, from a global perspective, Brazil not only is failing to catch up, it is falling relatively farther behind.

This chapter assesses Brazil's recent growth compared with other countries in Latin America and with several middle-income economies of approximately similar size. It then considers Brazil's rankings in a highly competitive global environment that is increasingly driven by knowledge and innovation. These international comparisons provide context for framing the issues of central concern for this report. This chapter also examines the structure of the Brazilian economy and its exports.

Chapter 2 presents a conceptual framework within which to interpret Brazil's experience, placing this study within a growing body of work on innovation, competitiveness, and economic growth. Chapter 3 looks more closely at the nature and origins of innovation. The remaining chapters focus on the specific innovation and human capital limitations that are constraining Brazil's current growth and competitiveness.

Brazil's Growth in Comparative Perspective

Between 1930 and 1980—approximately half a century—the Brazilian economy grew at an average rate of 7 percent per year. Indeed, during the latter years of that period, 1964 to 1980—often referred to as "the Brazilian miracle"—growth averaged a remarkable 7.8 percent. For about a decade during this period (1968–76, following the moderately successful stabilization

Julio Revilla and Carl Dahlman were key contributors to this chapter.

Figure 1.1. Inflation Rates, 1980–2007

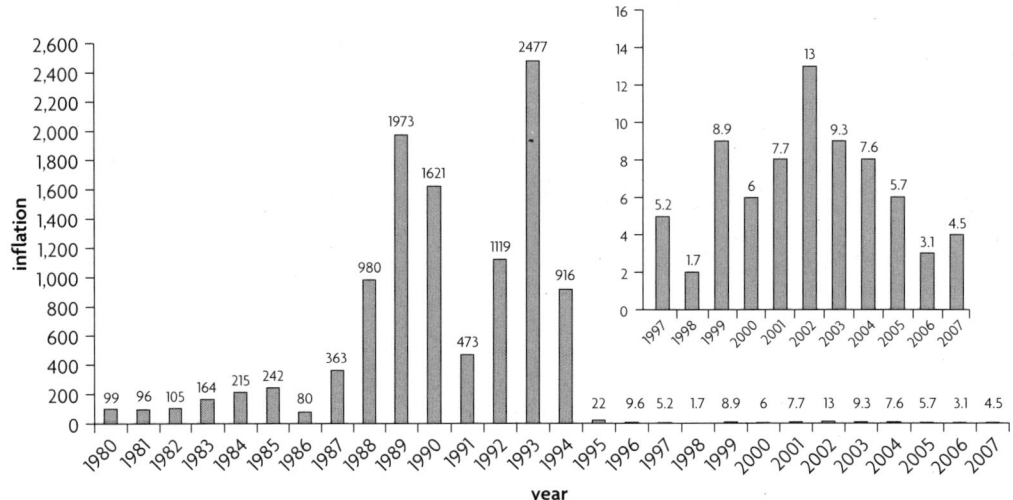

Source: IBGE (Brazilian Institute of Geography and Statistics), www.ibge.gov.br.

program that was undertaken in 1964), growth actually averaged about 10 percent annually.

Although growth was rapid during this period, the economy was not without problems. Volatility was moderate throughout and was strongly related to a series of external shocks and sharp policy reversals in economic policy.[1] As illustrated in figure 1.1, inflation was also high and was especially harsh in its impact upon the poor.

In the 1980s, GDP growth collapsed after the half century of sustained economic gain, and Brazil's economy may have experienced a long-term structural change. As shown in table 1.1, between 1981 and 1993 growth fell sharply—down to an average of 1.7 percent following the second oil shock of 1979 and Brazil's first debt crisis in 1981–82. This second period of recent economic history was marked by crisis and stagnation. It began with a steep output contraction and was marked by large macroeconomic imbalances and crisis-level output volatility—that is, very low growth accompanied by very large macroeconomic imbalances, high inflation, an external debt crisis, and repeated failures in stabilization efforts.

Brazil's most recent era, from about 1994 to today, was a period of limited recovery. It followed a stabilization program that was implemented under the Real Plan in 1994. As the stabilization program took hold and deepened, growth inched upward, achieving a positive but lackluster average rate of 2.8 percent between 1994 and 2005.

The overall picture of recent growth is captured in figure 1.2. The figure shows annual GDP growth as a percentage and as a 10-year moving average for 1964–2005. Although growth averaged nearly 10 percent in the 1960s and 1970s, it averaged only about 2.3 percent annually in the quarter century from 1981 to 2005.

Table 1.1. Average and Volatility of GDP Growth Rates, 1964–2005

	Average (%)	Standard deviation
"Brazilian Miracle," 1964–80	7.8	3.32
Crisis and stagnation, 1981–93	1.7	4.10
Limited recovery, 1994–2005	2.8	1.96

Sources: Based on the World Development Indicators (WDI) Database and data from the IPEA (Institute of Applied Economic Research), www.ipeadata.gov.br, and IBGE, www.ibge.gov.br, Web sites.

Figure 1.2. Annual GDP Growth: Percent and 10-Year Moving Average, 1964–2005

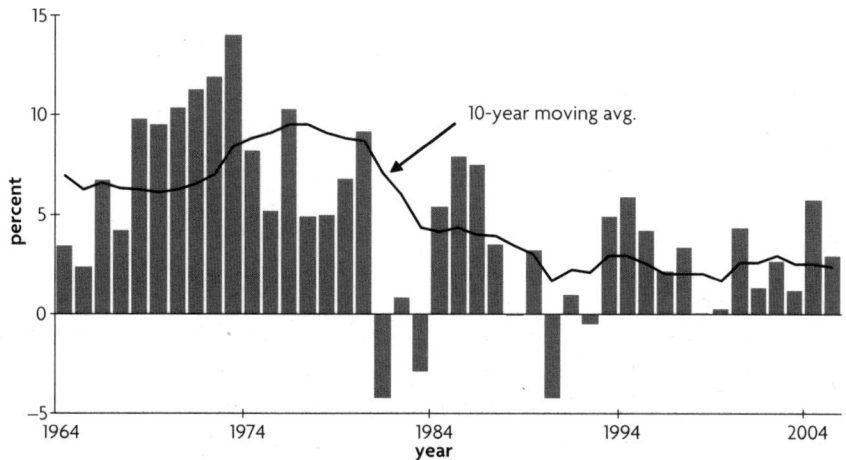

Source: Based on the WDI Database and data from the IPEA, www.ipeadata.gov.br, and IBGE, www.ibge.gov.br, Web sites.

Putting this picture in context, it is striking to note that both the high and low periods of growth (as well as the high volatility that accompanied the second and third periods) were almost completely out of sync with Brazil's regional neighbors as well as with other countries at similar levels of per capita income. As shown in table 1.2, Brazil's growth was significantly higher than the rest of Latin America during the 1960s and 1970s, but the situation was reversed in the 1990s, when Brazil's growth was lower than the rest of Latin America. This switch occurred even though the other economies were subject to essentially the same external environment; and ironically, many of them were highly dependent on Brazil.

The degree of relative underperformance is even more striking when the projected effects of macroeconomic stabilization and related policies are taken into account. In the early 1990s, most of Brazil's regional peers managed to bounce back from the so-called "lost decade." Brazil recovered gradually, but it hardly bounced back. Some of this failure might be explained by the 2001 Argentine contagion or by the 2002 Lula effect on higher interest rates. There was an apparently strong recovery in 2004; yet even so, it proved to be

Table 1.2. Annual Real GDP Growth Rate for Brazil and Select Countries
percent

	1960s	1970s	1980s	1990s	2000–05
Latin America	5.3	5.6	1.7	3.0	2.6
Argentina	4.1	2.9	−0.7	4.5	1.8
Brazil	5.9	8.5	3.0	1.7	3.0
Chile	4.4	2.5	4.4	6.4	4.4
Mexico	6.8	6.4	2.3	3.4	2.6
Asia					
China	3.0	7.4	9.7	10.0	9.3
India	4.0	2.9	5.9	5.7	6.4
Indonesia	3.7	7.8	6.4	4.8	4.7
Korea, Rep. of	8.3	8.3	7.7	6.3	5.2
East Asia	3.8	7.2	7.7	8.2	8.1
High-Income OECD	5.4	3.7	2.9	2.5	2.3

Sources: Based on the WDI Database and data from the IPEA, www.ipeadata.gov.br, and IBGE, www.ibge.gov.br, Web sites.

surprisingly short-lived. In 2005 and 2006, growth rose to about 3 percent, only slightly above the average for the previous decade.

The weakness in growth becomes even more apparent when Brazil's performance is compared with current fast-growth economies such as China, India, or Indonesia. As shown in table 1.2, during the 1960s these economies grew much more slowly than Brazil. Yet while Brazil fell flat during the 1980s, these economies managed to jump-start their growth. The Republic of Korea, Malaysia, and Thailand not only have sustained high growth for longer periods than Brazil but also have experienced prolonged periods of rapid expansion following periods of low growth.

The relative consequences of this low growth are illustrated in figure 1.3. As shown, Brazil's income gap relative to the Organisation for Economic Co-operation and Development (OECD) countries has steadily widened. Since the 1990s, Brazil has not only failed to catch up, it has fallen farther behind—from about 42 percent of OECD per capita income in 1980 to less than 29 percent in 2005.

Brazil's Competitiveness in an Increasingly Knowledge-Driven Global Environment

The generation of knowledge has significantly accelerated with the rapid advance of science and new communications technologies. Reductions in transportation costs, such as containerized shipping, are leading to the globalization of manufactured products, parts, and components, and the

Figure 1.3. Brazil's Per Capita Income Relative to the OECD Area (in PPP)

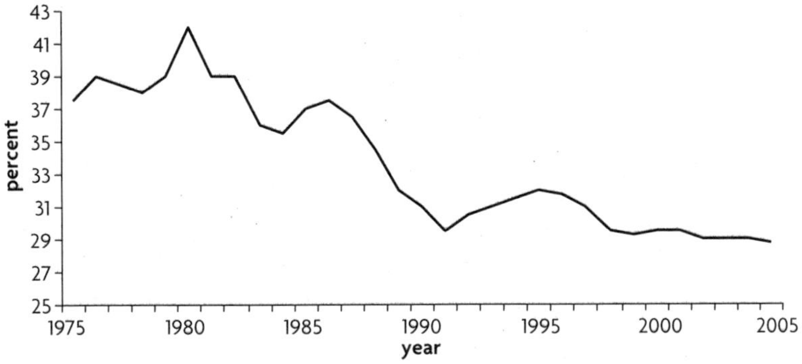

Sources: Calculated based on OECD data.

supply of inputs and raw materials from all around the world. The Internet, in particular, is making it possible to manage production facilities and trade globally in previously unimaginable ways. In addition, information and communications technologies (ICT) are permitting a growing trade in services; virtually any labor service that can be digitized is increasingly being outsourced and off-shored. Enhanced by trade liberalization, the decline in transportation and communication costs is leading to an increasingly globalized world. In the 15 years between 1990 and 2005, the share of imports and exports in global GDP increased from 38 percent to 55 percent.

In this new paradigm, it is knowledge—not natural resources or exports based on cheap labor—that constitutes the core of comparative advantage. As many cases have illustrated—including Bangalore, the capital of the Indian software industry—technical innovation and the competitive use of knowledge go hand-in-hand to produce high growth. Indeed, the proportion of goods in international trade with a medium-high or high technology content rose from 33 percent in 1976 to 54 percent by 1996 (World Bank 1999). Brazil was slow in adopting critical reforms that would have helped it ride the wave of this global shift to a knowledge economy. China, India, Korea, and most of the OECD countries advanced in making these critical reforms; and that is the main reason they outpaced Brazil.

As Porter (1990) and many subsequent authors have noted, comparative advantage among nations increasingly comes from technical innovation and the competitive use of knowledge—or from a combination of the two. Indeed, the real growth of value added in knowledge-based industries in many OECD member countries consistently outstripped overall growth rates during the past two decades. Growth of value added for the 1986–94 period was 3 percent for knowledge industries compared with 2.3 percent for the business sector as a whole (OECD 2000: 220, table 2).[2] Between 1985 and 1997, the share of knowledge-based industries in total value added rose from 51 to 59 percent in Germany, from 45 to 51 percent in the United Kingdom, and from 34 to 42 percent in Finland (OECD 2001).

One way to see the increased importance of knowledge is to examine the changing structure of global trade in even the short period between 1985 and 2004 (table 1.3).[3] The share of primary products decreased from 23.2 percent in 1985 to 14.7 percent in 2004, while that of manufactured products increased from 76.8 percent to 85.3 percent. This is largely because manufactured products are more income-elastic than primary products, because a greater number of new and increasingly differentiated products are produced through advances in knowledge. Moreover, the share of resource-based manufactured products fell from 19.4 percent to 15.6 percent. Low- and medium-technology manufactured products increased their share of global output by about 1 percent each. However, it was high-technology products—including electronics and airplanes—that made up for the decline in the share of primary and resource-based products. They increased their share from 11.6 percent to 22.4 percent. Thus, international competitiveness is now based much more on technological capability and innovation than on natural resources or basic production factors.

Reflecting the increasing importance of technology and innovation for competitiveness, the World Economic Forum (WEF) has devised a new Global Competitiveness Index (GCI) for 117 countries. The GCI separates countries into three stages of competitiveness development—factor-driven, efficiency-driven, and innovation-driven.

Figure 1.4 shows Brazil's rankings for a range of findings. Overall, Brazil ranks 57th out of 117 countries (the lower the ranking, the better the performance). On the basic requirements subindex (which characterizes factor-driven economies), Brazil ranks 77th. For component indexes, it ranks 79th for institutions, 70th for infrastructure, 91st for macroeconomy, and 52nd for health and basic education. On the efficiency enhancers subindex,[4] it ranks 51st for the component indexes, 50th on higher education

Table 1.3. The Changing Structure of World Exports, 1985 and 2004

Products	1985 (US$ billions)	2004 (US$ billions)	Annual growth rate (%)	1985 (%)	2004 (%)
All products	1,689	7,350	7.6	100.0	100.0
Primary products	391	1,018	4.9	23.2	14.7
Manufactured products	1,244	6,063	8.2	76.8	85.3
Resource based	327	1,148	6.5	19.4	15.6
Low technology	239	1,962	7.9	14.2	15.0
Medium technology	480	2,169	7.8	28.5	29.5
High technology	196	1,643	11.2	11.6	22.4

Source: CEPAL-TRADECAN 2005.

Figure 1.4. Brazil's Rankings on the Global Competitiveness Index, 2006

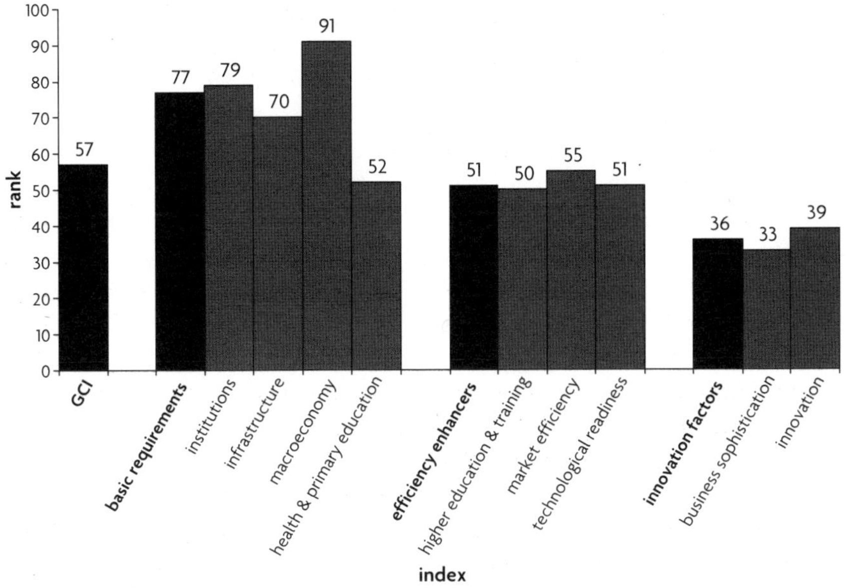

Source: World Economic Forum 2006.

and training, 55th on market efficiency, and 51st on technological readiness. Brazil ranks 36th on the innovation and sophistication factor subindex (which characterizes innovation-driven economies); it ranks 33rd on the business sophistication subcomponent and 39th on the innovation subcomponent.

The GCI scores suggest that Brazil will face a triple challenge if it wishes to make the transition from positive economic growth to rapid economic growth. First, it must improve upon the basic enabling conditions for growth—a sound macroeconomic environment, capable institutions, modern infrastructure, and higher-quality basic education and health services. Improving the basic enabling environment is probably the key priority because this is the area where Brazilian performance is the worst, particularly with regard to the macroeconomy. As shown in chapter 2, Brazil faces considerable obstacles in this area—largely as a result of low rates of investment—which negatively affect its ability to grow. Second, Brazil must improve domestic competition and market efficiency, education and training, and its ability to use existing technology effectively. Improving efficiency is the second key priority as it is the second-worst performance area; chapter 2 examines some of the problems with market efficiency in greater detail. Third, Brazil must improve its capacity to undertake innovation through business sophistication and the ability to develop, adopt, and disseminate new products and processes. The rankings show that Brazil does relatively better in this area than in the other two. However, looking to the future, this is an increasingly important area because of the importance of knowledge and innovation for competitiveness.

Finland provides a good example of how knowledge can be a force to drive economic growth and transformation. During the 1990s, Finland became the economy most specialized in ICT in the world, completing its transition from an economy based on natural resource exploitation to one driven by knowledge and innovation. Export diversification has been integral to Finland's improved economic performance. This diversification was attributable largely to continuous emphasis on tertiary education, linkages and spillovers among industries, and new knowledge-based enterprises. Since 1980, investment in research and development (R&D)—primarily by the private sector, with the government as an important secondary partner—has more than doubled. R&D investment reached the equivalent of 3.5 percent of GDP in 2004, far above the European Union (EU) average of less than 2 percent. The Finnish innovation system also has succeeded in converting its R&D investments and educational capacity into industrial and export strengths in the high-technology sectors (Dahlman et al. 2005).

A new type of enterprise—producer-services companies providing specialized information in support of manufacturing firms—has recently begun to emerge. These companies are a principal source of created comparative advantage and value added among the highly industrialized economies (Gibbons 1998). In the knowledge economy, advances in microelectronics, multimedia, and telecommunications give rise to important productivity gains in many sectors. They are also the key to a multitude of new products in a wide range of new industrial and service activities. On the down side, the ever-faster creation and dissemination of knowledge means that the life span of technologies is becoming progressively shorter. Obsolescence sets in ever more quickly.

Developing economies are often affected by these transformations without experiencing the benefits. The capacity to harness knowledge for sustainable development and higher living standards is not equally shared. In 1996, it was estimated that OECD countries accounted for 85 percent of total investment in R&D; Brazil, China, India, and the newly industrialized countries of East Asia accounted for 11 percent—and the rest of the world, only 4 percent. One reason agriculture is so much more productive in industrial countries than it is in developing countries is that the former spend up to five times more on agriculture-related R&D than do the latter. In other words, industrial countries possess the combined infrastructure, expertise, organizational arrangements, and incentive structures to allow their R&D investments to become productive. The exclusive group of advanced economies enjoys a virtuous circle in which the benefits of research help to produce the wealth and public support that perpetuate their ability to continue investigation on the scientific frontier (Romer 1990).

Figure 1.5 compares the economic evolution of Brazil and Korea from 1958 to 1990. The figure well illustrates the dramatically different outcomes for two countries, both of which started with roughly similar GDP per capita— but one of which adopted a knowledge-based development strategy. The graph is based on the standard Solow method of accounting for economic growth. It represents a stylized attempt to estimate the relative contribution of tangible factors—such as the accumulation of physical capital and additional

Figure 1.5. Knowledge as a Factor in Income Differences between Brazil and the Republic of Korea, 1956–90

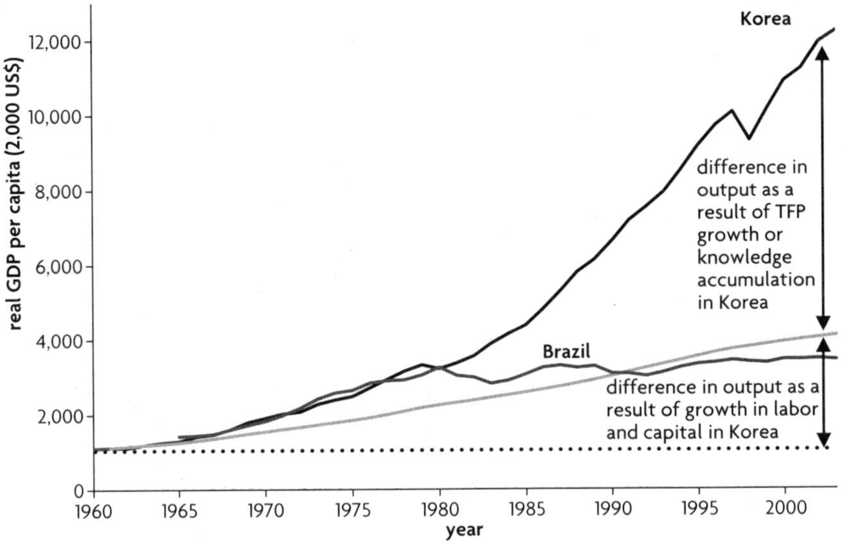

Source: Calculations based on World Bank internal data. Knowledge for Development (K4D) Program, World Bank Institute.

years of schooling in the labor force—and factors linked to the use of knowledge, such as the quality of education, the strength of institutions, the ease of communicating and disseminating technical information, and the level of management and organizational skills (Solow 2001). In this model, technical progress raises the potential output from a given set of inputs. Empirical measures are then applied to assess the extent to which growth is attributable to increased inputs (more labor and capital) or to the use of inputs in a more productive way. The latter measure, commonly referred to as total factor productivity (TFP), is closely linked to how knowledge is used in production. Because TFP measures output per units of input, raising it may lead to higher standards of living.

The differing growth trajectories illustrated in the figure reflect a broadly observed pattern, not just circumstantial differences unique to Brazil and Korea. Easterly and Levine (2000) analyzed several similar cross-country growth studies, and they also concluded that differences in TFP growth are the main explanation for differences in economic growth. Accordingly, they argue for a shift in policy emphasis to focus on TFP rather than simple capital accumulation.

The Structure of the Economy and the Structure of Exports

Two structural elements of the Brazilian economy that affect the country's growth and competitiveness are worth highlighting. The first is that Brazil—like other Latin American economies but in contrast to rapidly growing economies like China and India—has experienced relatively little structural change in the composition of economic activity over the past 25 years. By 1985, Brazil and the other key Latin American economies had already made the major transition from agriculture to industry. This occurred in the past 25 years for China and India, which transitioned from agriculture into industry and services. The shift from low-productivity agriculture to higher-productivity industry (or services) helps to increase overall growth; and it is one of the reasons for the faster growth of China and India.

The service sector can be a very important source of growth. As seen in table 1.4, India's recent growth rates of over 8 percent have been led by knowledge-intensive services. While the share of services in GDP expanded slightly in Brazil, it is 6 percentage points below the average of 60 percent for middle-income economies and the average of 65 percent for high-income economies. This is due to neglect of the service sector in Brazil's development strategy, even though services account for more than half of GDP. The growth potential of the service sector is especially significant because it is rapidly becoming the largest knowledge-intensive sector of economic activity.[5] For OECD countries, the share of medium- and high-technology manufacturing value added in total economic activity averages only 7.5 percent; however, the average share of knowledge-intensive market services is 20 percent.[6] Thus, Brazil needs to do much more to realize the potential of its service sector; and, as is argued below, doing so depends on improvements not only in the business environment but also in educational attainment and quality.

The changing structure of exports over the past 25 years is also revealing when comparing Brazil with other countries. As shown in table 1.5, Argentina is still primarily an exporter of food and fuels, although there has been an increase in the latter at the expense of the former, and manufactured exports have increased from a quarter to a third. Chile is still primarily an exporter of ores (particularly copper) and food, and manufactures have only increased 5 percentage points to 14 percent. In Brazil, there has been a reduction

Table 1.4. Changing Structure of Output between 1980 and 2005, Selected Countries

	GDP (US$ billions)		Agriculture (%)		Industry (%)		Manufacturing (%)		Services (%)	
	1980	2005	1980	2005	1980	2005	1980	2005	1980	2005
Argentina	77	183	6	9	41	36	29	23	52	55
Brazil	235	796	11	8	44	38	33	—	45	54
Chile	28	115	7	6	37	47	21	18	55	48
Mexico	195	768	8	4	33	26	22	18	59	70
China	202	2234	30	13	49	48	41	34	21	40
India	172	805	38	18	26	27	18	16	36	54

Sources: WDI 1998 and 2007.
Note: — = not available.

Table 1.5. Changing Structure of Merchandise Exports between 1980 and 2005
percent

	Food		Agricultural raw materials		Fuels		Ore and metals		Manufactures	
	1980	2005	1980	2005	1980	2005	1980	2005	1980	2005
Argentina	65	47	6	1	3	16	2	3	23	31
Brazil	46	26	4	4	2	6	9	10	37	54
Chile	15	19	10	7	1	2	64	56	9	14
Mexico	12	5	2	1	67	15	6	2	12	77
China	—	3	—	1	—	2	—	2	—	92
India	28	9	5	2	0	11	7	7	59	70

Source: WDI Database.
Note: — = not available.

Table 1.6. Exports by Technology Intensity, 2004
percent distribution

	Argentina	Brazil	Chile	Mexico	China	India
Natural resources	51.4	32.6	41.5	14.6	3.2	15.6
Resource-based manufactures	24.5	21.9	49.2	6.4	6.9	29.8
Low-technology manufactures	7.4	11.0	2.1	13.5	39.2	35.5
Medium-technology manufactures	14.1	24.9	5.5	37.5	19.0	12.8
High-technology manufactures	1.7	7.9	0.5	24.2	30.5	5.4
Other	0.9	1.7	1.2	3.8	1.1	0.9
Total	100	100	100	100	100	100

Source: CEPAL-TRADECAN 2005.

of 20 percentage points in the share of food. Most of that decrease has been made up by an increase in the share of manufactures from 37 percent to 54 percent. However, the share of manufactures in Brazil's total merchandise exports appears relatively small when compared with 70 percent for India, 77 percent for Mexico (where the share of fuels plummeted from 67 percent to 15 percent as the difference was more than taken up by manufactures), and 92 percent for China.

Table 1.6, which uses the same classification as that used in table 1.3, shows that Brazil is still relatively specialized in exports of natural resources and natural-resource-based manufactures (55 percent of the total), and very weak on high-technology manufactures (7.9 percent, compared with 24.2 percent for Mexico and 30.5 percent for China). The world's average for exports of high-technology manufactures against all exports is 29 percent.

Table 1.7 shows that Brazil has a revealed comparative advantage only in natural resources and natural-resource-based manufactures and some simple labor-intensive manufactures (food and beverages)—all the items above the line. In addition, Brazil has lost comparative advantage in most manufactured products except machinery and transport equipment, wood and cork, non-metallic minerals, and oils and lubricants. The improvement in machinery and transport equipment (where, nonetheless, Brazil still does not show a real comparative advantage) is due to its exports of truck chasses and airplanes.

Overall, what is happening on the export side is that Brazil is continuing to specialize in natural resources and natural-resource-dependent manufactures. This is part of a broader global picture. China's entry into the global trading system in a major way appears to be having three major impacts on the world—and on Brazil. First, China's tremendous competitiveness in manufactured goods (China is already the world's third-largest exporter of merchandise exports) is helping to drive down the cost of manufactured products. Second, because of its voracious appetite for natural resources

Table 1.7. Brazil's Revealed Comparative Advantage (RCA), 1995 vs. 2005

Product	% of BR exports	Revealed CA	
	2005	1995	2005
Crude materials, inedible	16.04	3.35	5.49
Leather manufactures	1.39	3.04	4.33
Food and live animals	18.82	3.03	3.85
Animal and vegetable oils and fats	1.29	4.97	3.59
Wood and cork manufactures	1.41	1.89	2.73
Iron and steel	7.81	3.11	2.51
Beverages and tobacco	1.53	2.42	1.72
Manufactured goods classified chiefly by material	18.90	1.53	1.34
Rubber manufactures, nes	0.94	1.52	1.27
Non-ferrous metals	2.33	2.03	1.24
Paper, paperboard, and manufactures	1.29	1.40	0.89
Non-metallic mineral manufactures	1.47	0.73	0.73
Machinery and transport equipment	26.39	0.49	**0.67**
Mineral fuels, lubricants	6.11	0.15	**0.61**
Chemicals	6.83	0.69	0.60
Textile yarn, fabrics, made-up articles	1.14	0.66	0.55
Manufactures of metal, nes	1.12	0.67	0.55
Miscellaneous manufactured articles	4.09	0.47	0.34
Commodities & transactions not classified, accumulated	0.00	0.49	0.00

Source: Calculated from World Integrated Trade Solutions/UNTRANS.
Note: Revealed comparative advantage is share of sector in Brazilian exports/share of sector in world exports. Bold represents products showing an increase in RCA; gray represents products showing a decrease in RCA over the decade.

and commodities, China is driving up its prices on the global market. In fact, China has reversed the famous negative terms of trade against natural resources made famous by Raul Prebisch.

Third, the decrease in the cost of manufactured products represents a windfall for global consumers, Brazilians included (except for the still relatively high tariff and nontariff barriers to imports). The lower consumer prices are caused partially by China's rapid expansion of low-cost manufactures, and partially by rapid global advances in technology and innovation. The increased demand for natural resources and commodities represents a boon for natural resource and commodity exporters. This has benefited Argentina, Brazil, and Chile. Producers in these sectors are making record profits. As with other natural-resource- and commodity-exporting countries, Brazil's increased export revenues, as well as foreign direct investment (FDI) inflows to these sectors, are leading to an appreciation of the currency, which is in turn causing so-called Dutch disease.

While these two impacts are positive for Brazil, the third impact is negative. The increased competitive pressure from Chinese manufactured products is causing many producers of manufactured products—in Brazil and other developing countries, such as Mexico—to close. Some Brazilian producers in the shoe and textile sectors are closing down their production facilities in Brazil and are contracting production in China.

Conclusion

In summary, Brazil is benefiting from the global boom in demand for natural resources and commodities; and it should continue to do so. Brazil has been successful in applying knowledge to leverage its agricultural resources by investing in agricultural R&D (which has raised productivity in wheat and soy, which has helped to boost exports); and it has developed its ethanol program to substitute for high petroleum prices. Brazil should continue to invest in knowledge to leverage the return from its natural and agricultural resources. In the short run, it has to improve the broader enabling environment, particularly to reduce the very high cost of capital and the cost of doing business. It will also have to address the growing overvaluation of its exchange rate as a result of this commodity boom.

It is also clear from this preliminary analysis that Brazil is not making sufficient use of knowledge that already exists abroad, or even in the country. This is in part a result of a poor enabling environment—in particular, a restrictive trade policy that denies Brazilian firms access to better inputs (especially capital goods at world prices). This point is elaborated on in chapter 4. In addition, the low investment rate impedes the upgrading of production through the introduction of technology embodied in more advanced capital goods. Thus, trade policy issues must also be addressed in the short run.

In the medium and longer run, care must be taken to avoid overspecialization in natural and agricultural resources. Booms in commodity prices come and go. The current boom is likely to continue so long as China continues its rapid growth; however, like other booms, this one will eventually bust. Brazil must maintain its competitiveness in many manufacturing sectors by improving its technological and innovation capability across the board.

Besides making more effective use of existing knowledge (the shorter-term agenda identified above), Brazil must do better at investing in new knowledge-intensive sectors that may have greater future growth potential. Currently, Brazil is not receiving economic returns commensurate with its investment in R&D. Improving the efficiency of these investments will require better allocation and management of existing resources, as well as more investment by both the public and the private sector in the medium and longer term.

As is argued later in this report, Brazil's capacity to effectively assimilate and use existing knowledge—no less than its capacity to create knowledge or invent new technologies on the frontier—depends on the breadth of educational attainment and the acquisition of basic skills within the workforce. Addressing the highly unsatisfactory state of basic education needs to be carried out in parallel with addressing the higher-level technological requirements for global innovation. The short-term issues have to do with better allocation of existing resources. The longer-term issues have to do with sustained educational investments that will improve the quality of education from primary through postgraduate levels. Chapter 2 continues the analysis of Brazil's low growth and places innovation and education in this broader context.

CHAPTER 2

Behind Brazil's Slow Growth

This chapter lays out the broad conceptual framework for this report. It begins with the traditional neoclassical growth conceptualization in which output is understood as a function of capital, labor, and technical change. The present research builds upon this traditional growth accounting model with the explicit addition of innovation and enabling environment, thereby creating a four-factor schema—enabling environment (used here as synonymous with investment climate), physical capital, human capital, and TFP (used here as synonymous with innovation).

The chapter briefly explains how each of these four factors is related to growth and, more specifically, Brazil's underperformance in growth in recent decades. The conceptual framework provides a broader context for our later focus: innovation and education (referring here to a process of human capital formation). Chapter 3 expands upon the concept of innovation. Subsequent chapters address these components of innovation at the macro (national) and micro (firm) levels and then discuss human capital formation as it affects innovation and competitiveness in Brazil.

Conceptualizing Growth and Deriving a Revised Model

In the conventional neoclassical model, growth generally is understood to be a function of capital and labor, with technology treated largely as a given. In endogenous growth theory, change is treated as something that happens within the model itself—in other words, technology is factored in. In the first conceptualizations of growth-accounting models, any part of the growth output that could not be attributed to capital and labor was attributed to technological change (equated with innovation). That is to say,

---

Julio Revilla and Carl Dahlman were key contributors to this chapter.

Output = Function of Capital + Labor; and
Change in Output = Function of change in Capital + change in Labor
+ Technological change residual

The technological change residual often has been referred to as "the residual of our ignorance"—the problem-solving "mystery variable" that would explain why economies such as Brazil's and Korea's, which had roughly similar endowments of capital and labor 30 years ago, subsequently grew at such different rates (see figure 1.5 in chapter 1). As research has deepened into this process, the variable has solidified and taken shape as TFP growth. TFP can be understood as the factors beyond capital and labor that enable an economy to increase production output. While the classic factors of capital and labor remain critical in any explanatory conceptualization of output growth, TFP increasingly is seen to be the real driver within economies. Indeed some studies suggest that TFP accounts for as much as 60 percent of economic growth within some countries. Moreover, as this chapter shows, Brazil's slow growth in the past decade is attributed partly to stagnant productivity, whose levels are influenced heavily by TFP.

Identifying the factors that compose TFP is difficult. Many elements—ranging from better intermediate inputs to improved organization and management, as well as large-scale, new, or improved technology—can increase TFP. This report focuses primarily on the innovation component of TFP, including the creation and use of knowledge that is new, the acquisition of existing (foreign) knowledge, and the use of existing (in Brazil) knowledge in new or more efficient ways. These aspects of innovation are developed more fully in chapter 3.

A substantial literature has developed around the new approach to growth factors that is taken here. Many models adjust for input quality. Capital, for example, is typically refined and measured in terms of capacity utilization, or sometimes in terms of equipment vintage. Labor has been refined and measured through education, skills, and experience. The more that capital and labor are adjusted to account for knowledge components, the lower the residual of technical change. Empirical estimates of the contribution of knowledge or innovation to growth therefore depend on how much the components have been adjusted for knowledge-related factors. Some recent models have also begun explicit consideration of innovation-related variables such as R&D, patents, foreign investment, and technology licensing.

In addition, some models have begun to incorporate the investment climate. While investment climate might be correctly thought of as a subcategory of "enabling environment," the constellation of macroeconomic, regulatory, and governance regimes—the structures and forces that shape investment decisions—is used here as largely synonymous with the broader term.

In the conceptual model that guides this report, TFP (that is, innovation) is an explicit, endogenous factor. Our conceptual model is represented schematically in figure 2.1 as a four-box framework. As shown by one-way and reciprocal arrows, growth is the interactive result of physical capital, TFP (innovation), and human capital, with interaction strongly defined by an overarching enabling environment that can either enhance or obstruct it.

Figure 2.1. A Conceptual Model for the Components of Growth

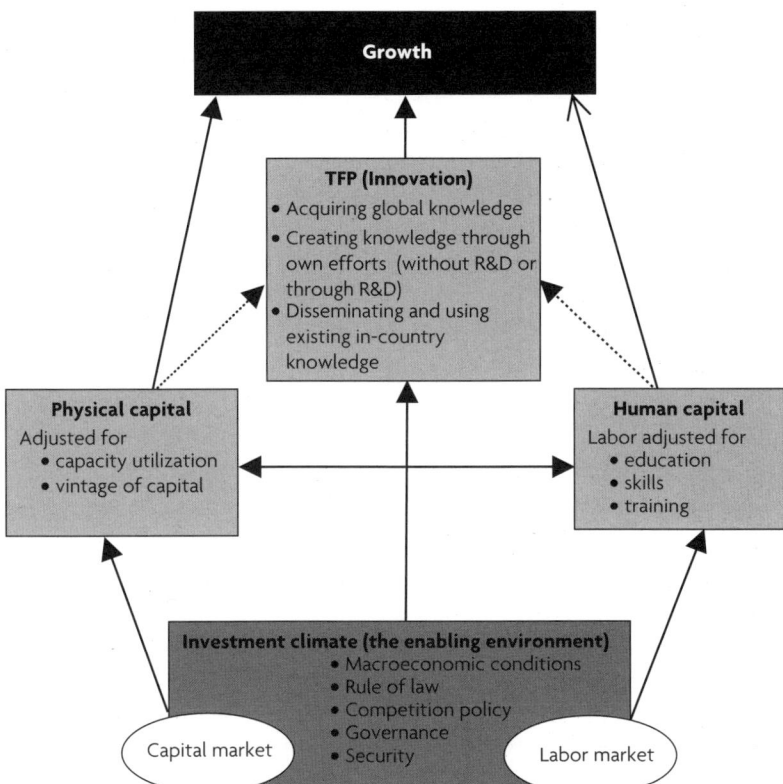

Source: Authors.

In the schematic representation, notice that physical capital is made operational and measurable as capacity utilization and capital vintage (specifically plants and equipment). Similarly, human capital is made operational and measurable as the education, skills, and training that are added to labor. Finally, within innovation, the model distinguishes between knowledge creation through autonomous innovative effort (either *through* R&D or *without* R&D), acquisition of foreign knowledge and technology, and wider dissemination and application within a country of knowledge and technology already possessed by that country (see chapter 3). Clearly, strong interactive effects link all the variables, which can make it particularly hard to isolate the contribution of any single factor. Given that practical difficulty, econometric analysis rarely takes the interactions of all factors fully into account.

Figure 2.1 should be understood as a broad schematic representing the aggregate level of a country. Because the complexity of a complete econometric analysis was beyond the immediate purposes of this report, advanced analytic work at the macro level was not undertaken. Instead, we briefly summarize other work and use the findings in subsequent sections of this report to sketch out what occurs within each of the four factors. However, to deepen the analysis, econometric work at the firm level was conducted. This was based on the 2003 Investment Climate Assessment of Brazil and draws upon significant recent work by Brazilian researchers. These findings are presented more fully in chapter 5.

The next sections of this chapter look at the relationship between the elements of the four-box framework and growth. The first section discusses the relationship between innovation and economic growth. Subsequent sections explore relationships between physical capital and economic growth, and between human capital and economic growth.

Innovation and Economic Growth

The literature was surveyed initially to assess the linkage between innovation indicators and economic performance. Surprisingly, this topic has received scant attention. Therefore, our focus shifted toward factors that *determine* innovation. The problem, in short, was not the lack of analysis on innovation but that innovation has typically been analyzed as a determinant of productivity growth rather than of overall economic growth (for example, see Grilliches 1990; or Jaffe and Trajtenberg 2002). Therefore our analysis gravitated toward Lederman and Saenz (2005), one of the few studies that examines the effect of innovation on long-term development.

Lederman and Saenz employ input measures that include patent activity, R&D expenditures, participation of engineers and scientists in R&D activity, and the public-private makeup of R&D. Their analysis then links these variables with level of development (GDP per capita), an increasingly popular dependent variable in the literature concerned with growth rates. The study finds evidence that innovation's effect on development is as large (or larger) than the associated effect of the "rule of law," another variable that has received much recent attention. Empirically, the innovation environment is shown to exert an apparently strong and direct effect on development. This result is robust to different specifications, including instrumental variable specification.

In translating their econometric evidence to actual country performance, Lederman and Saenz take particular note of China and India. Both of these rapid-growth economies have invested heavily in R&D, with India relying more on publicly financed R&D and China relying more on acquiring technology developed elsewhere. By contrast, Latin America and the Caribbean clearly lag behind, and the investment gaps caused by this lag are important in explaining the relative differences in economic growth.

In reviewing the literature, a problem in robustness of analysis became clear. In general, too much was expected of formal R&D and patenting in developing countries. Because developing countries are behind the global frontier, we decided to use a broader definition of innovation that does not narrowly focus on R&D, patenting, and the creation of new products. We also considered knowledge that may be new to the country, or even the firm.

To better understand the dynamics of the process, supporting evidence was sought linking firm-level innovation with economic growth. The literature indeed contained factors that influence innovation at the micro level (for example, Souitaris 2002). Our own econometric analysis, presented in chapter 5, discusses some of these factors in greater depth. This perspective is important because, as a practical matter, it sheds light on the kinds of environments that foster innovative behavior, especially on the role of the national innovation system and the role of skills and education. The chapter 5 firm-level evidence demonstrates intriguing links between certain forms of innovation and economic growth—for example, between discoveries and new products.

How does innovation improve productivity that leads to economic growth? Total factor productivity provides useful clues. As previously noted, TFP attempts to explain why one economy would perform better than another given similar capital and labor inputs from traditional growth accounting (see, for example, Solow 1956). Rather than changes in *factors* (such as total investment or population growth), TFP focuses on changes in *productivity* related to improvements in education, training, and technology, among others. Because this focus on processes departs from the neoclassical assumption of exogenously determined technological change, TFP is a derivative of so-called new growth theory.

A simple way to conceptualize TFP empirically is as the "Solow residual," that is, the part of the economic growth production function that is otherwise unexplainable. This is a common way to operationalize TFP. For example, Country A's endowment of investment and labor might be expected to produce a certain level of growth. However, when actual growth departs from the expected path, the difference (that is, the residual) is commonly attributed to unobserved factors related to productivity—in other words, TFP.

The most obvious problem with this formulation is that it emphasizes precisely that part of the growth accounting model that cannot be explained directly. A related problem is the disentanglement of productivity effects from factor effects. One way to handle these issues, albeit imperfectly, is to compute the TFP contribution to growth and then regress this parameter onto other variables in a multivariate regression. Significant linkages are thus established between TFP and viable explanatory variables. This helps to tighten the analysis of the mechanisms that explain TFP and also helps to reinforce the validity of the concept. Take for example the analysis of TFP in Sub-Saharan Africa by Akinlo (2005). Among other things, this study finds that secondary education enrollments are associated positively with TFP, while factors such as external debt are negative predictors.

Compared with other countries, how does Brazil fare in the strength of TFP or in the factors that can strengthen it? The most obvious comparison is with East Asia. An interesting debate pits those who are skeptical that TFP is behind the East Asian Miracle (most notably, Paul Krugman 1994) and those who argue that hidden productivity factors played a significant role (Singh and Trieu 1996). Using case studies, Singh and Trieu find evidence that as much as half the growth from 1965 to 1990 in Korea, Japan, and Taiwan was due to TFP. They make several comparisons with Latin America and conclude that TFP is a significant factor in the East Asian Tigers' much better performance.

Two key points stand out from the TFP literature. First, TFP is a conceptually important tool for understanding how innovation stimulates productivity and thereby economic growth. At the very least, the notion of TFP provides a plausible starting point for explaining why some countries grow faster than others. This has significant policy implications because it underscores that a country's endowment does not rigidly determine its growth, and that choices matter. Second, the conceptual nature of TFP is consistent with the causal chain laid out in figure 2.1—namely, that the broader enabling environment affects not just the rate of investment or human capital accumulation, but also the efficiency with which all factors are used.

The Relationship of Physical and Human Capital with Economic Growth

For several theoretical growth models—in which the initial values of human capital and per capita GDP matter for subsequent growth rates—physical capital accumulation is viewed as one source of economic growth. The main insights about the effect of capital accumulation on growth stem from Solow, the founder of the neoclassical growth model. In this model, the rate of technological progress is assumed to be constant, and the growth process is entirely accumulation-driven (Helpman 2004). In endogenous growth models, per capita growth and the investment-to-GDP ratio tend to show a positive relationship.[1] In other models that include human capital, an increase in the initial stock of human capital tends to raise the ratio of physical investment to GDP.[2]

Empirical evidence largely supports the relationship between physical capital and economic growth. Baier et al. (2006) examined the relative importance of the growth of physical and human capital and the growth of TFP. They used data on 145 countries that varied in the starting year but all ended in 2000.[3] The authors found that, over long periods of time, the growth of output per worker is associated with the accumulation of physical and human capital as well as with technological change. For all countries, weighted average results showed that output per worker grew 1.61 percent per year while physical capital, human capital, and TFP per worker increased 2.33, 0.92, and 0.22 percent per year, respectively. Results are similar for Latin America, and for Brazil in particular, as shown in table 2.1. Overall, the authors conclude that TFP growth is a somewhat important part of average output growth per

Table 2.1. Average Growth of Output and Inputs
weighted average

Region/country	Growth rate per worker				TFP growth relative to output growth
	Output	Physical capital	Human capital	TFP	
All countries	1.61	2.33	0.92	0.22	0.14
Latin America	1.59	2.27	0.86	0.26	0.17
Brazil	1.67	2.18	0.67	0.50	29.97

Source: Baier et al. 2006.

worker, but the largest share of change can be attributed to growth of aggregate input per worker. This conclusion is similar to that of Jones (2002), who used a model of idea growth to explain economic growth. Jones found that for the United States, the deepening of physical capital, the increase in educational attainment, and the rise in R&D intensity accounted for 81 percent of U.S. economic growth from 1950 to 1993.

In general, Latin America experienced much lower growth rates over the past 25 years than did East Asian countries. Keeping in mind that Brazil, on average, grew more slowly than the rest of Latin America, it may therefore be instructive to compare Latin America[4] with East Asia, South Asia, and comparators such as Eastern Europe and Sub-Saharan Africa (figure 2.2).

The lower growth in Latin America in figure 2.2 can be explained by three factors. First, the lower rates of capital growth are associated with much lower levels of savings and investment. These in turn are related to poorer macroeconomic management and to generally poorer investment climates, particularly where the productive sectors are subject to international competition and higher costs of capital. Second, TFP growth has been significantly lower in Latin America than in East Asia. Essentially, this TFP gap reflects weaker innovation systems. The consequence of this gap was a shackling of growth potential in Latin America, and in Brazil in particular. Third, the contribution of human capital (which in this exercise was treated separately from the labor input) was lower. As shown in the figure, the contribution of human capital was lower in Latin America not only compared with East Asia, but also compared with South Asia and Sub-Saharan Africa.

Another growth exercise was performed using data for Latin America and developed countries. The data covered 30 countries from 1950–92. Results, to be interpreted as systematic time effects corresponding to the extension of the lag of each variable on GDP per capita, showed that annual increments of 1 percent in physical capital in the short term of five years would increase

Figure 2.2. Growth and TFP—Latin America Compared with Other Regions

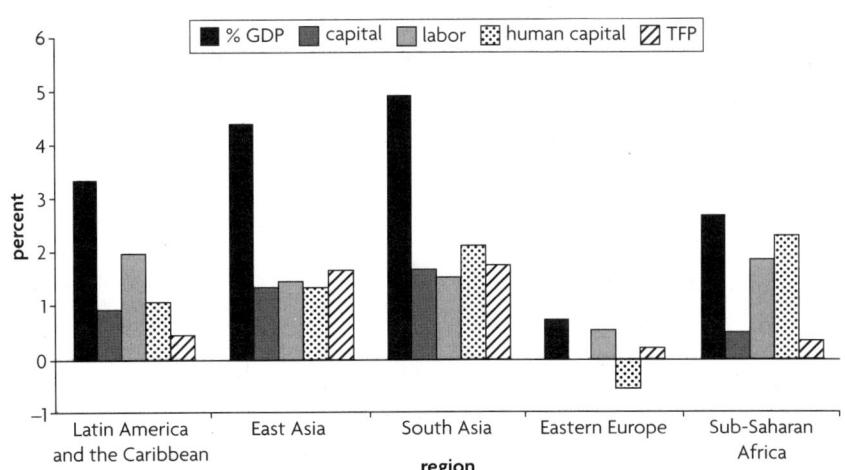

Source: IDB 2006.

GDP per capita by 2.96 percent in developed countries, 0.94 percent in Brazil, 0.76 percent in Central America and in the group of Andean countries, and 0.66 percent in the group composed of Argentina, Chile, Colombia, Mexico, and the República Bolivariana de Venezuela. For developed economies, physical capital, technology, and government size have a significant effect on GDP per capita. There is also strong evidence that human capital has a greater impact on physical capital (Arraes and Teles 2003).

There is certainly evidence, based on cross-country regressions, that having a more educated workforce leads to higher growth (Barro 1996). But these conclusions have come under attack both on methodological (Levine and Renelt 1992) and substantive (Pritchett 1996) grounds. However, new research that focuses cross-country analysis on quality rather than quantity of education is resuscitating confidence in the macroeconomic relationship between human capital and economic growth (Hanushek and Wößmann 2007). In addition, evidence suggests that R&D spending and the stock of scientists in a particular society co-vary with economic performance (Meyer et al. 2000). Finally, there are also the enabling institutional conditions for productivity-enhancing innovation, such as the rule of law and intellectual property right protection.

Looking at the link between training and productivity and growth, some empirical studies have shown both a positive interaction between education and training (Blundell et al. 1999) and positive returns of training to the individual, the firm, and overall economic growth (Blundell et al. 1999; Bartel 2000). In interesting work with cross-country OECD data, Coulombe, Tremblay, and Marchand (2004) showed that differences in average literacy skill level explain 55 percent of the differences in the long-term per capita GDP growth in 14 OECD countries. Even more interesting, based on data from the past 45 years, a 1 percent incremental increase in the average literacy of a given nation would return a 1.5 percent permanent increase in GDP per capita and a 2.5 percent increase in productivity. A disaggregation of this OECD data provides useful conclusions that could well apply to Brazil in the future: the percentage of individuals with high skills appears to have little positive impact on long-term growth in OECD economies. In contrast, the percentage of individuals with very low literacy skills exerts a strong negative effect on growth. In sum, basic skills for the entire population appear as important to growth as the development of sophisticated, high-level skills within a country.

A Growth Decomposition Exercise

In the early 1990s, Brazil adopted an orthodox macroeconomic policy framework that encompassed fiscal discipline, a floating exchange rate, and inflation targeting. According to most observers, these were the right things to do. Yet even though they may have been right, they do not appear to have been enough. Although Brazil recovered stability, it did not achieve rapid growth. For that, something was missing. To understand what, it is helpful to look comparatively at the three recent economic periods from the perspective of the conceptual model discussed above.

The slowdown during the 1980s was significantly associated with drastic declines in capital formation and productivity. Employment growth played a much less important role. As shown in table 2.2, gross capital formation fell from its near 10 percent average during the first period, 1964 to 1980, into relative stagnation during the second period, 1981 to 1993. This decline mirrored the pattern of GDP growth. During the limited recovery period from 1994 to 2005, capital accumulation, and especially productivity, bounced back somewhat. Gross capital formation rose at an annual average rate of 2.6 percent, while employment growth declined. This suggests that the growth of the past 25 years strongly reflects declining capital accumulation and an associated decline in productivity.

A simple growth decomposition exercise for the three recent periods (table 2.3) confirms and extends this picture. Obviously, the results of growth decomposition depend on the parameters chosen; however, for a common range of values, after accounting for capital (column A) and employment (column B), decomposing the contribution to GDP growth shows that the residual—TFP, which we have loosely equated with technological change—emerges as a highly important factor in accounting for Brazil's performance decline (column C). Using a capital elasticity of 0.5 (the most common figure found in cross-country studies), TFP growth declined from 1.32 percent during the first period to 0.16 percent during the second period, before inching up to 0.50 percent during the third period. With an elasticity of 0.3, the basic picture remains the same: capital growth explains most of the change in growth rates throughout the three periods.

Table 2.2. Growth of GDP, Capital Stock, and Employment, 1964–2005
percent per year

	GDP	Gross capital formation	Employment
"Brazilian Miracle," 1964–80	7.8	9.9	3.11
Crisis and stagnation, 1981–93	1.7	−0.3	3.39
Limited recovery, 1994–2005	2.8	2.6	2.06

Sources: Based on the World Development Indicators (WDI) Database and data from the IPEA, www.ipeadata.gov.br, and IBGE, www.ibge.gov.br, Web sites.
Note: Gross fixed capital formation for capital, economically active population for employment. Estimates for 2005.

Table 2.3. Contribution to GDP Growth, 1964–2005

	Gross capital formation (A)	Employment (B)	TFP (C)
"Brazilian Miracle," 1964–80	4.96	1.55	1.32
Crisis and stagnation, 1981–93	−0.14	1.70	0.16
Limited recovery, 1994–2005	1.30	1.03	0.50

Sources: Based on the WDI Database and data from the IPEA, www.ipeadata.gov.br, and IBGE, www.ibge.gov.br, Web sites.
Note: Gross fixed capital formation for capital, economically active population for employment. Estimates for elasticity of capital (α) = 0.5.

These findings are consistent with other estimates of TFP calculations for Brazil. Pioneer studies include those of Elias (1992) and De Gregorio (1992). Fajnzylber and Lederman (1999) and Loayza et al. (2004) have provided extensive reviews on Latin America. Detailed analysis of the Brazilian experience has been provided by Gomes et al. (2003), Pinheiro et al. (2004), and the World Bank (2005a).[5] In a recent study, Ferreira et al. (2006) show that Brazil, among other Latin American economies, had periods in the 1960s and 1970s in which its TFP was even higher than TFP estimates for the United States. Yet as shown in table 2.4, Brazil's TFP relative to that of the United States dropped from 1.07 in 1975 and 1.02 in 1980 to 0.8 in 1995 and 0.73 in 2000.

The lower growth rate in Brazil for the crisis and stagnation period (1981–93) compared with the "miracle" period (1964–80) was caused by negative growth in gross capital formation (table 2.2), as well as nearly flat TFP (table 2.3). The direct contribution of labor to growth did not change greatly (except for a small decline in the limited recovery period from 1995 to 2005). As discussed below, a low rate of gross capital formation is a reflection of the overall investment climate (the "enabling environment" for growth), which is similarly affected by macroeconomic instability, high interest rates, a weak regulatory regime (leading, for example, to labor-market problems), and poor rule of law. In terms of our conceptual framework, a correspondingly low rate of TFP has several related causes—low investment (because much technical change is embodied in new equipment), a poor investment climate, and underinvestment in education and skills.

The TFP estimates in table 2.3 and table 2.4 suggest that past improvements in productivity apparently took place during periods of capital expansion in Brazil, which is to say that technological progress was achieved through the acquisition of new capital. However, this characterization leaves important questions unanswered from a policy perspective. After the seemingly successful macroeconomic stabilization and structural reforms that started in 1994, why did Brazil *not* return to the high growth levels of the 1970s? If the post-Real stabilization plan was indeed successful, why isn't Brazil growing faster? Does the relatively modest 2.5 percent average between 1996 and 2006 actually represent a new ceiling for Brazil?

Table 2.4. Relative TFP of Brazil and Latin American Comparators
U.S. = 100

	1960	1965	1970	1975	1980	1985	1990	1995	2000
Brazil	83	80	88	107	102	86	75	80	73
Argentina	99	93	93	98	93	75	58	74	69
Chile	68	64	73	64	76	65	72	87	80
Colombia	81	80	90	91	96	87	90	77	64
Mexico	109	111	110	118	113	99	79	74	77
Latin America	87	86	89	93	88	75	68	69	62

Source: Ferreira et al. 2006.

Several recent econometric exercises have suggested what Brazil's output growth could be, at least in principle. But these estimates basically confirm that something is happening that is preventing current growth from again reaching the high levels of the 1960s and 1970s.

Investment Climate—The Enabling Environment for Growth

As noted in the conceptual framework presented in figure 2.1, the investment climate has a major effect on growth, as it does on the three other components—physical capital, human capital, and TFP. What are the key elements of the investment climate that are negatively affecting growth in Brazil?

High Taxes, Large Government, and Poor Expenditure Quality

Government spending exceeds 40 percent of GDP, and tax revenue rose to 38.8 percent of GDP in 2006. The high tax burden discourages private investment, formal sector employment, and economic growth. The Brazilian tax system is among the most burdensome in the world, because of both its high rates and its administrative complexity. On average, the tax burden represents nearly 150 percent of gross profits, compared with an average of 53 percent in Latin America as a whole. Growth is also constrained by the composition of public spending, currently characterized by low investment rates and high government consumption (mostly in salaries and social security pensions). The relatively large public sector debt (and interest payments) is reinforced by rising primary spending, low investment, and budget rigidities. In addition, the quality of government spending in Brazil suffers from weak public sector management and institutional arrangements (World Bank 2007a).

Table 2.5 shows the tax that a medium-size company must pay (or withhold) in a given year, as well as indices of the administrative burden in paying taxes. These measures include the number of payments entrepreneurs must make; the number of hours spent preparing, filing, and paying; and the percentage of profits they must pay in taxes.

Table 2.5. "Doing Business" in Comparative Perspective

Indicators	Brazil	Latin America	OECD
Payments (number)	23.0	41.3	15.3
Time (hours)	2,600	431.0	203
Profit tax (%)	22.4	22.8	20.7
Labor tax and contributions (%)	42.1	14.5	23.7
Other taxes (%)	7.2	11.8	3.5
Total tax rate (% profit)	71.7	49.1	47.8

Source: World Bank 2006b.

High Interest Rates

Average lending rates remain high in Brazil—around 50 percent in real terms in 2005, among the world's highest rates—despite significant financial system reforms during the 1990s. Barriers to more efficient financial intermediation include the large size of governmental borrowing, directed credit schemes that account for about a third of total bank lending, and less-than-efficient public banks. High interest rates and financial intermediation spreads—around 38 percent in real terms in 2005—are explained by the high levels of public debt and by issues of jurisdictional uncertainty that weaken creditor rights. Insufficient creditor protection because of flaws in the legal system and juridical practice also contribute to costly financial intermediation and high spreads (World Bank 2004b, 2004d, and 2006a).

Lack of Infrastructure Investment[6]

A significant share of fiscal adjustment has involved cuts to public infrastructure investment, which have had significant negative impacts on investment and growth at the firm level. Total public investment fell from about 5 percent of GDP in the 1980s to about 2 percent of GDP in 2002–05. Fiscal adjustment has relied on cutting infrastructure investment in part because of rigidities in current spending. Private sector investments in infrastructure have not compensated for the reduced public investment. Indeed, private infrastructure investment has fallen, except for sales of government shares and concession rights in the privatization of telecommunications, electricity, transport, and (to a lesser extent) water and sanitation during 1992–2001. Despite recent changes to the regulatory environment, private infrastructure development has been hindered by the lack of a stable and credible regulatory environment and the lack of improving cost recovery by investors. Provision of infrastructure services is an important aspect of logistics costs (the so-called *Custo Brasil*), which is estimated at about 20 percent of GDP. These services include transport, warehousing, inventory, and customs—all of which are affected negatively not only by inadequate infrastructure but also by interest rates and red tape (World Bank 2004d; and World Bank 2006).

Inadequate Labor Institutions and Legislation

The Brazilian labor market is affected significantly by relatively inflexible labor legislation. Labor institutions and the labor law are in continuous interplay through a forest of regulations that often lead to low labor productivity and low growth in formal sector employment. Labor legislation in Brazil is heavily geared toward job security, resulting in low formal sector employment and productivity growth and higher growth in the informal labor market. Paradoxically, the Brazilian labor market shows a very high turnover because large severance payments mandated by law induce worker dismissals before seniority triggers take effect. This results not only in litigation but in low-productivity jobs, because firms have less incentive to train workers. Consequently, labor productivity has been affected negatively. Labor markets, institutions, and regulations in Brazil reinforce income inequalities because greater informality, fewer incentives for job training, and high payroll taxes all encourage labor-market informality (World Bank 2002a; and World Bank 2005a).

Improving but Limited Trade Openness

In the 1990s, trade policy was significantly reformed. Compared with the 1970s and 1980s, tariffs on imports were substantially reduced. Nevertheless, Brazil remains a relatively closed economy—international trade accounted for about 30 percent of GDP in 2005. Although exports have grown strongly as a share of GDP (from 10.7 percent in 2000 to 16.8 percent in 2005), this process has been driven largely by higher international commodity prices. Some industrial sectors indeed have become more competitive. Not surprisingly, evidence suggests that some low-productivity industrial sectors are precisely those that face the least foreign competition. Tariffs, nontariff trade barriers, and administrative barriers to investment remain as significant comparative drags. Although trade has opened in Brazil, other countries have opened at an even faster pace. This reluctance to open trade has limited the positive effects of increased competition at a global scale. The limited effect of Brazil's greater openness in trade also seems related to the faltering of two other complementary policies: investment in human capital and labor-market reform (World Bank 2002a and b; and World Bank 2004e). Box 2.1 below presents Brazil's trade regime, compared with other countries.

Box 2.1. Brazil's Trade Regime Compared with Other Countries

Two indicators of trade barriers show that Brazil, in spite of some liberalization over the last 10 years, continues to have a relatively protected trade regime.

The table below is based on an index developed by the Heritage Foundation, which includes not only average tariffs but also nontariff barriers such as quotas, quantitative restrictions, labeling, and licensing requirements. It shows no progress on trade liberalization since 1995 (the first year for which the index was calculated). In addition, the table shows that (with the exception of India) Brazil continues to have a more restrictive trade regime than the average Latin American country, China, or the United States.

Index of Tariff and Nontariff Barriers

	Brazil	Argentina	Chile	Mexico	LAC	China	India	United States	Western Europe
1995	3.50	4.00	4.00	2.50	3.95	5.00	5.00	2.50	2.50
2006	3.50	3.00	1.50	2.50	3.00	3.00	5.00	2.00	2.03

Source: Knowledge Assessment Methodology (KAM) 2007.
Note: A score is assigned to each country based on the analysis of its tariff and nontariff barriers to trade, such as import bans and quotas, as well as strict labeling and licensing requirements. Based on the Heritage Foundation's Trade Policy Index, the score takes on values from 0 (most favorable) to 5 (least favorable).

(continued)

Box 2.1. (*continued*)

The second table shows that, except for India, Brazil's mean and weighted mean tariffs are higher than for China, the United States, and the OECD, and are higher than the average for low- and middle-income developing countries, or even for Latin America. Disaggregation for primary versus manufacturing products, however, shows that Brazil's tariffs on manufactured products continue to be higher than for the other countries except India; and the average U.S. and OECD tariffs continue to be lower overall. Brazil's mean and weighted mean tariffs on primary products are lower than those of China and India, and are lower than the average for low- and middle-income developing countries and Latin America. Thus, Brazil is still quite protectionist in its manufacturing sector.

Tariff Barriers
percent

	Brazil	China[a]	India	United States[b]	OECD	Low & middle income	LAC
Mean tariff							
1990	31.8	42.9	81.8	6.3	—	—	—
2005	12.3	9.2	17.0	3.2	3.1	9.4	9.6
Mean weighted							
1990	33.0	40.6	83.0	4.4	—	—	—
2005	7.1	4.9	14.5	1.6	2.0	6.1	5.3
Primary products mean tariff							
1990	25.7	36.2	74.1	4.5	—	—	—
2005	7.9	8.8	24.4	2.8	3.7	12.3	11.9
Primary products mean weighted							
1990	13.1	22.3	49.5	2.4	—	—	—
2005	1.5	3.4	16.5	0.8	2.1	5.9	3.9
Manufactured products mean tariff							
1990	33.2	44.9	84.1	6.7	—	—	—
2005	12.6	9.2	15.9	3.3	3.0	9.0	9.3
Manufactured products weighted tariff							
1990	39.4	46.5	93.6	4.8	—	—	—
2005	9.2	5.3	12.8	1.8	1.9	6.1	5.6

Source: WDI Database.

Note: The mean tariff is generally considered a better general indicator of protection than the weighted mean. The latter often biases the rates downward because higher tariffs may discourage imports and reduce the weights applied to these tariffs. However, sometimes imports of commodities with higher tariffs are still made and the weighted mean may be higher. Therefore, both mean and weighted mean tariff rates are presented.

a. Data for China are for 1992 and 2005.

b. Data for the United States are for 1989 and 2005.

Judicial System Inefficiency

The judicial system in Brazil has an unenviable track record for slowness, unpredictability, and inefficiency. The complexity of the system, the time required to reach decisions, and the overall costs of contract enforcement greatly undermine contract efficiency. Arbitration procedures exist but are rarely used at the outset. Instead, courts are typically used by one party to force the other into arbitration. Trials are lengthy, and multiple appeals are common.[7] As a result, the judicial system is unusually overloaded. The Brazilian Supreme Court, for example, handles more than 100,000 cases a year, compared with about 200 cases handled each year by the U.S. Supreme Court. Although courts play an integral part in the aforementioned problems, the judicial system also includes the property registries, government lawyers, and the Public Ministry or Attorney General's Office (World Bank 2004d).

Red Tape

The size and cumbersome structure of Brazil's three levels of government— federal, state, and municipal—clearly impose burdens on business operations. On average, starting a business, registering property, and paying taxes in Brazil require substantially more time and are more expensive than elsewhere in Latin America, and are far more burdensome than in other regions. Some Brazilian states have started to simplify procedures for registering businesses, for example, through the creation of one-stop shops. By and large, however, the overall process remains costly and slow, with the most time required in São Paulo, where a remarkable 152 days is required to register a business. Across states, registering a property takes, on average, 61 days, placing Brazil 17th out of 22 countries in Latin America. Complex entry and property registration procedures, as well as high taxes, have another downside. The large number of time-consuming, costly procedures not only hinders business entry, but also lays the foundation for and encourages corruption (World Bank 2006). Table 2.5, above, presents some of this evidence.

Trade Orientation, the Export Sector, and Growth

Many studies of growth have found an important relationship between trade orientation (which is part of the broader incentive regime in the enabling environment) and exports. The acceleration of growth is often linked to export expansion, especially from the industrial sector. Bonelli (1992) studied the relationship among TFP, output growth, and trade orientation for the 1975– 85 period preceding trade liberalization. Sectoral data for manufacturing and extractive industries and a comparison between the two quinquennia permit an interesting analysis of macroeconomic performance in light of corresponding policies. As might be expected, Bonelli finds a positive association between export expansion and rates of productivity change as estimated by TFP growth. Export expansion followed a program of trade liberalization that was launched in 1979. Despite the larger crisis that then enveloped the economy, the short-lived program of export expansion contributed substantially to the growth of nearly all industries from 1980 to 1985.

Periods of increased TFP (and corresponding growth) can also be linked to lower import tariffs that effectively reduced protection for domestic industries but coincided with productivity gains for the sector overall. Ferreira and Rossi (2003) provide empirical analysis on how trade liberalization that began in the 1980s affected industrial sector productivity growth. By analyzing the periods before and after trade liberalization, they show that TFP grows faster at lower rates of protection. The findings are less conclusive for countries such as Argentina, Chile, and Mexico. But for Brazil at least, a strong case can be made that trade liberalization had a positive impact on TFP and growth.

Moreira (2004) examined the relationship between trade liberalization and increased productivity, and also concluded that liberalization leads to stronger growth. His estimates suggest that the productivity increases following Brazil's trade liberalization in 1988–90 were actually larger than those in Mexico following the North American Free Trade Agreement. He attributes subsequent slow growth to the lack of an aggressive trade policy. The result was a disproportionate distribution of benefits. The positive effects of liberalization on productivity were concentrated in the relatively small export sector rather than distributed across the economy more broadly. This underlines the need for institutional reforms and consolidation of macroeconomic stability in order to expand the export sector.

Does the Public Sector Constrain or Catalyze Growth?

Many observers over the past two decades have pointed to the large size of Brazil's public sector as a growth constraint, particularly as it affects both the cost of capital and high taxes.

From 1950 to 1980—a period of high growth and boom—the public sector was the main agent for investment and the chief catalyst for growth in Brazil. However, with the fiscal weakening and debt crisis of 1982, the government's capacity to invest was reduced substantially. At the same time, the private sector investment was unable to fill the gap, in part because it was held back by high interest rates and high taxes, related in turn to the large size of the government sector.

Explanations differ as to why the Brazilian economy slowed so dramatically in the 1980s and failed to recover its previous dynamism.[8] Yet there is a growing consensus that the size of the government has been—and continues to be—an important factor. Using consolidated tax revenues as a simple proxy for size of government, Brazil has the largest government (relative to GDP) among large middle-income economies (including Argentina, China, India, Mexico, and Russia), and it has a larger government than economies that have entered the high-income category.

Why does this matter? The significant increase in government consumption and the corresponding contraction in public and private investment are at the core of both the TFP and growth problems. The exceedingly large public sector results in high taxes, high interest rates, and lower infrastructure investment, all of which impede efficient resource allocation (especially in the use of technology) and, hence, growth.

To analyze comparable figures of government size, we look at the relative size of government consumption (to eliminate investment) in figure 2.3. The first column shows that since the 1988 constitution (when government spending began to rise substantially), Brazil nearly doubled government consumption as a percentage of GDP. In contrast, government consumption rose modestly in comparator countries such as China, India, and Korea.

Three reasons have frequently been cited to explain the dramatic slowdown in growth after 1980—the large surge in government consumption (figure 2.3), sharp increases in the relative price of investment (Bacha and Bonelli 2004), and high vulnerability to international liquidity (Barbosa 2001). It can be argued that all three are related to the size of the public sector. The large share of government consumption contributes to a low level of savings and, hence, investment. The increase over time of the relative price of investment (capital goods) in Brazil has also been linked to greater government intervention through higher distortions. Vulnerability to international liquidity (or external conditions) emerged as a major issue in the financial crisis that affected emerging markets from the 1980s to early 2000. In the case of Brazil, this was mostly due to a sizeable increase in external liabilities, mostly by the public sector.

Adrogué et al. (2006) demonstrate empirically that the steady rise in government consumption since the mid-1980s has negatively affected per capita growth. Loayza et al. (2004) and Bacha and Bonelli (2004), among other researchers, have demonstrated the same. Most empirical models show that macroeconomic stability efforts normally correlate with improved growth—including stabilization of the debt-to-GDP ratio, a successful inflation targeting

Figure 2.3. Government Consumption as a Percentage of GDP in Four Countries

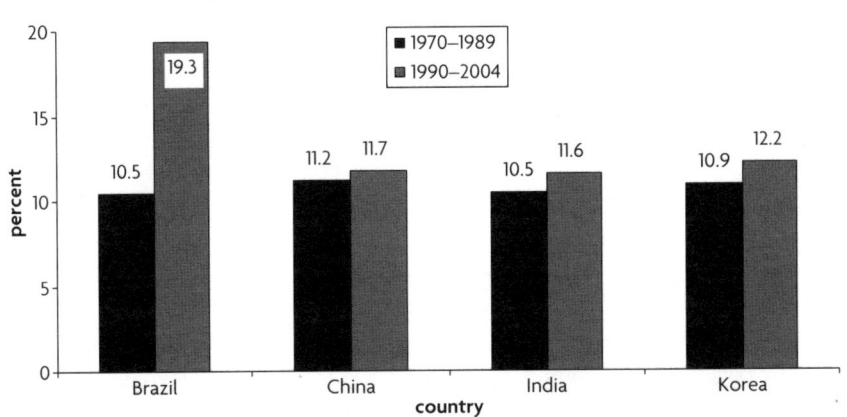

Sources: Based on the WDI Database and data from the IPEA, www.ipeadata.gov.br, and IBGE, www.ibge.gov.br, Web sites.

regime, flexible exchange rate, and most other structural reforms implemented in the 1990s. Despite its successes in these areas, Brazil's growth performance was disappointing, particularly when compared with previous periods or with international competitors. Despite efforts on the fiscal front, public debt remains large, and more significantly, real interest rates remain high (at about 10 percent in real terms for the central bank policy rate).

Although large, Brazil's public debt—at about 45 percent of GDP in net terms and about 66 percent of GDP in gross terms—is not significantly different from that of middle-income countries, such as India, the Philippines, and Turkey. And if the total public debt is below other middle-income countries that have faster growth rates, why are interest rates so high in Brazil?

Hypotheses include the following: market uncertainty over true public sector liabilities (for example, remaining skeletons from indexation, ballooning social security commitments); judicial and property-rights-related uncertainty; and lack of competition or poor regulation within the financial sector. High public sector consumption (the Brazilian government has become a net spender since the 1980s) is a leading factor in the relatively low level of savings and investment. This helps to explain why the intertemporal price of consumption, namely the real interest rate, is so high. A large government burden from high consumption (and low savings and investment) is interrelated with high taxes and high interest rates.

Large government consumption also negatively affects government investment in infrastructure. Gomes et al. (2003) and Adrogué et al. (2006) show empirically that despite all the efforts on the macroeconomic front, the sharp reduction in government investment during the 1990s and after 2000 has been a major factor in disappointing growth rates. Weak telecommunications, poor roads, inefficient ports, unreliable air transportation, questions on energy sustainability, and unequal access to water are all obvious obstacles to strong trade, commerce, and business.

Conclusion: Getting the Fundamentals Right Was Necessary but Insufficient

Following nearly a decade of economic turmoil in the 1980s, Brazil adopted an orthodox macroeconomic policy framework that encompassed fiscal discipline, a floating exchange rate, and inflation-targeting. Subsequently, Brazil successfully stabilized inflation and its exchange rate, and it is starting to reduce its public debt-to-GDP ratio. The fruits of tight policy were positive growth, but not rapid growth.

Fiscal policy in the aftermath of the Real Plan helped Brazil to reduce its public debt-to-GDP ratio and to increase the sustainability of public debt (mostly though increased tax revenues). Monetary policy based on an inflation-targeting framework and a flexible exchange-rate regime reduced inflation from 12.5 percent in 2002 to 3.1 percent in 2006 while cutting foreign-exchange risks. Greatly improved debt management helped to drastically slash the external public debt. Good fiscal policy was helped by a highly favorable external environment in which strong export growth has generated current account surpluses since 2003.

Getting the macroeconomic economic fundamentals right (even if they still fall far short of being "perfectly right")—was enough to achieve moderate positive growth in the range of about 3 percent a year. Yet this level did not keep Brazil from falling increasingly behind its global competitors much less allow it to catch up and overtake them.

The current Brazilian government has announced plans to increase government infrastructure spending to increase productivity and growth. In doing so, the government is constrained by its high current spending and the ongoing debt burden that limits borrowing and, therefore, spending capacity. The continuous growth in government size during the past decade—with rising tax burden for the private sector and constrained domestic savings—does little to encourage the private sector to take up the slack in investment. As the historical evidence in this chapter indicates, even though productivity in Brazil improved during the past decade, it remains less than that of earlier periods when investment grew faster.

With our conceptual model of growth now defined and linked to the existing literature, it is time to apply it to various aspects of the Brazilian experience. Chapter 3 takes a closer look at a key element—innovation. Chapter 4 assesses the macro outcomes and the institutional and legal framework of innovation in Brazil, including all its forms—creation, acquisition, adoption, dissemination, and use of knowledge and technology. The same broad definition of innovation is then applied at the firm level (chapter 5). Chapter 6 assesses and analyzes the contribution of human capital in Brazil, covering basic education and basic skills development as well as tertiary education and advanced skills development.

CHAPTER 3

Defining Innovation

What Is Innovation?

This and the two chapters that follow develop the concept of the national innovation system in Brazil and look more closely at the relationship between innovation and growth at the national (chapter 4) and firm (chapter 5) levels. We begin with a fundamental question—what is innovation? Innovation—defined broadly to include products, processes, and new business or organizational models—is not just advancement upon the frontier of global knowledge but also the first use and adaptation of global technology in new contexts. Therefore, this study examines both the role of R&D in creating new knowledge and also the process through which this new knowledge is "commercialized" and translated into more rapid economic growth at the national or firm level. Because much new technological knowledge is acquired from abroad, this study looks at the various means through which foreign knowledge can be captured and adapted to the local context. Finally, it suggests that the wider dissemination and more creative use of existing in-country knowledge can be a critical step in increasing productivity in Brazil.

This chapter looks at the three sources of innovation—creating, acquiring, and using new knowledge. In our conceptual framework (chapter 2), we emphasized that innovation alone—like enabling conditions or physical capital alone—is insufficient to generate rapid economic growth. Technology by itself provides no magic. For new knowledge to translate into TFP-driven growth, something more is needed. Productive workers are the missing link—not only highly trained scientists who can be called upon to invent something new, but shop-floor workers who can be called upon to do something new. For this reason, Brazil as a nation will be called upon to "innovate" in how it educates the 45 million young people who are enrolled in

---

Carl Dahlman was a key contributor to this chapter.

its school system. Chapter 6 explores the formation of human capital—at the primary, secondary, and tertiary levels—in terms of the three sources of innovation discussed here. At the end of this chapter, a table displays the three sources of innovation as a typology that provides an integrated view of the policies, instruments, and institutions of the national innovation system in Brazil. This lays a foundation for the recommendations to be presented in chapters 7 and 8.

The Creation and Commercialization of Knowledge

The creation of knowledge is usually associated with inventive activity, especially the creation of new technology. Innovation in this sense (especially in Brazil) typically brings to mind scientists working in universities and engineers working in R&D labs. Figure 3.1 shows global R&D efforts for Brazil and other comparator countries in purchasing power parity (PPP) terms.

Innovation is by no means limited to formal R&D efforts—not all R&D results in invention and not all invention comes from formal R&D. On the contrary, invention and knowledge creation may be produced by constantly trying to improve hands-on production—or for that matter, through accident, serendipity, trial and error, and sometimes sheer luck.

The process of invention is frequently so idiosyncratic and nonlinear that investment in creating new technological knowledge is particularly difficult and risky. No one knows in advance what level of national investment is likely to produce what level of innovation, much less at what point in the process something commercially useful is likely to emerge. Invention is just the first step in innovation. Theoretical discoveries in basic knowledge are often first

Figure 3.1. R&D Effort for Brazil and 11 Comparators

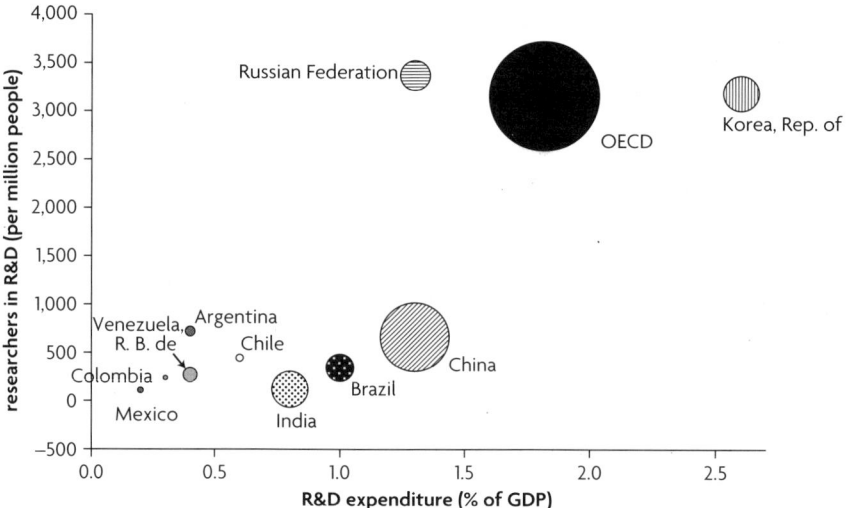

Source: Knowledge for Development (K4D) program, http://go.worldbank.org/AW9KZWJB10.
Note: The size of the circle represents each country's total R&D expenditure for 2003 (PPP, current international dollars).

published in scientific and technical journals. Figure 3.2 compares Brazil's output of scientific and technical journal articles compared with the outputs of advanced and neighboring countries.

If an idea or insight is sufficiently unique, it may be patented—at which point it might spawn an entire new industry or, more likely, never be used at

Figure 3.2a. Scientific and Technical Journal Articles per 100,000 Inhabitants (Other Countries)

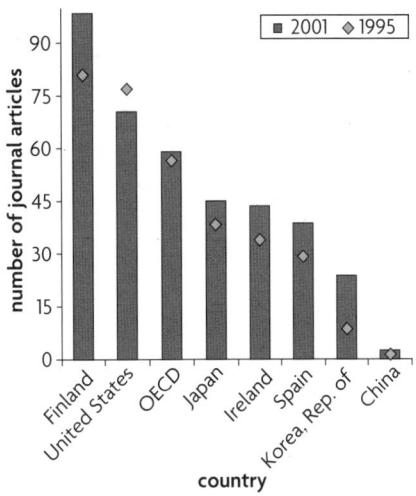

Source: IDB 2006.
Note: The scale for figure 3.2a differs from the scale for figure 3.2b.

Figure 3.2b. Scientific and Technical Journal Articles per 100,000 Inhabitants (Latin America and the Caribbean)

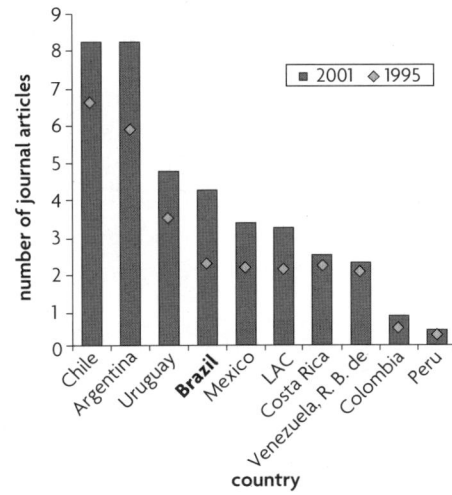

Source: IDB 2006.
Note: The scale for figure 3.2a differs from the scale for figure 3.2b.

Figure 3.3. Patents Granted by the U.S. Patent and Trademark Office to Brazil and Regional Comparators

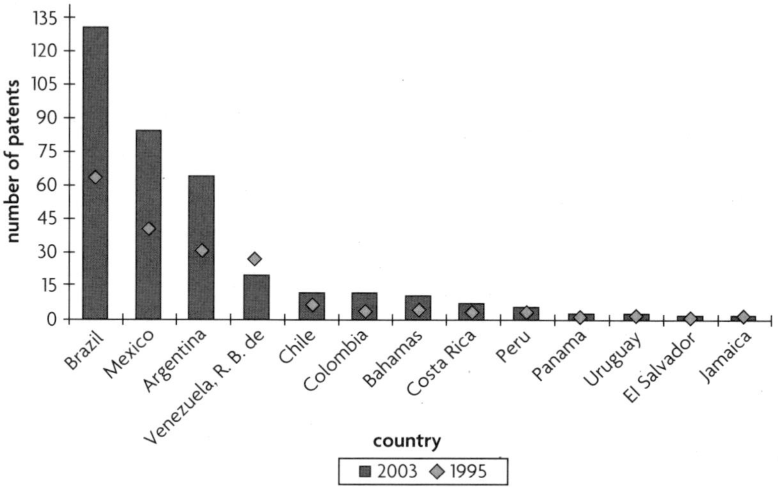

Source: IDB 2006.

all (see figure 3.3). New knowledge also may be kept as a trade secret once patented. Almost always, further development and engineering work (and costly investment, extending through much iteration) is required to convert the discovery into a practical application. Finding a concrete marketable application often costs far more than the original invention, and in many cases the process exceeds the capacity or interest of the patent holder. For this reason, innovation tends to show up on economists' screens only when it reaches the point of commercial application.

The first ever application of an innovation can be thought of as a global innovation. The first use of knowledge where it has not been used previously is considered to be an innovation over prevailing local practice. Because developing countries are well behind the global technological frontier, they generally obtain far greater economic benefit from using knowledge that already exists than from trying to create new knowledge. This is not to say that developing countries should not try to create new knowledge, which may have many indirect positive externalities in addition to the direct economic benefits described here; but from an economic viewpoint, it is generally more efficient for developing countries to acquire and use new knowledge than it is to create new knowledge.

In Brazil, public R&D labs, universities, and some productive enterprises are the main incubators of "created knowledge" and thus constitute a fundamental part of the national innovation system. Although government and university research labs are typically the main actors, productive enterprises are the most important segment because they are the main appliers of new knowledge. At the global level, multinational corporations typically drive the creation and dissemination of applied knowledge. It is estimated that transnational companies carry out more than half of all global R&D. Indeed, R&D budgets of many large multinationals frequently dwarf the total R&D expenditures of all but the largest developing countries. In 2002, for example, the U.S. car manufacturer General Motors spent US$5.4 billion on R&D, almost $1 billion more than Brazil's total R&D spending that year. (See UNCTAD 2005.)

If benefits are to accrue from new knowledge in the national innovation system, public laboratories, universities, and private firms must interact and cross-fertilize each other. As emphasized in chapter 7 (recommendations), this interaction must be fostered by instruments and mechanisms tailored for that purpose. For example, government grants can require research participation by more than one kind of actor in an R&D process. Similarly, subsidies can be provided that facilitate the exchange of personnel between labs, universities, and firms.

If new knowledge is to be used economically, attention must be paid to the process and prospects for commercialization. It is not ignoble or venal to think about patents and commercialization as basic research is conceptualized and undertaken—especially in a country such as Brazil, where the creation of new technical knowledge is financed primarily through public expenditure. In terms of public policy, this translates into mechanisms ranging from tax incentives to science parks. "Incubators" are needed to encourage interaction between publicly funded scientists and the private sector, and to ensure that this interaction benefits society at large. Where scientists lack experience or business acumen, mechanisms are needed that provide social benefits by translating ideas into viable enterprises. Publicly funded incubators can play a wide range of roles, from matching scientists with businesspeople who can help develop business plans, get permits, find employees, and obtain financing for start-up operations. Support of this sort would not in itself be entirely "innovative." Many of Brazil's key competitors in the global marketplace—countries that once lagged far behind but are steadily forging far ahead—already are doing precisely that.

Acquiring Foreign Knowledge

Creating new knowledge is far riskier and requires more technological capability than acquiring new technology. A country (or firm) needs to know not only what is relevant, but what is worth negotiating for and at what cost over the long run. Acquiring foreign knowledge also requires significant technological capacity, including research infrastructure. Universal primary education has nearly been achieved, and universal access to secondary education is imaginable on the horizon, so Brazil has a large and potentially productive population base with which to build an innovation economy. The problem is that the educational system has not yet oriented itself toward meeting the challenge. Improving educational quality and human capital formation across the board (not just for a few elite scientists) represents, of course, an enormous human challenge, but it is also Brazil's most significant economic opportunity for recapturing the high economic growth rate it once enjoyed.

There are many means to hasten the acquisition of necessary technology: direct foreign investment; licensing; technical assistance; technology embodied in capital goods, components, or products; copying and reverse engineering; foreign study; technical information in printed or electronic form (including what can be accessed through the Internet); twinning; training arrangements; and others. Much relevant technology is already in the public domain or is owned by governments that potentially can place it in the public domain. In the case of proprietary technology, which by definition must be sold or transferred on a contractual basis, gaining access can be more complex. The fact is, legitimately or not, proprietary technology almost always "leaks," depending on the capability of users and on the intellectual property rights (IPR) regime governing a contractual transfer. And while intellectual property rights are fundamental to the creation of new knowledge and technology, IPR regimes do change, and public policy has considerable sway over the "rules of the game." Therefore, countries that contract and use proprietary technology should be fully prepared to capitalize on legitimate opportunities for knowledge transfer when they arise. In short, for both public and proprietary technology, there are ample opportunities for the transfer of usable knowledge to an innovation-ready population interested and intellectually equipped to put it to productive use.

Disseminating and Using Knowledge

Once new technology is acquired, it is disseminated primarily through commercial activity—through sale and transfer, as well as through imitation and replication by "copycat" consumers, enterprises, and organizations. Like the process through which knowledge is selectively acquired (see above), technology is disseminated primarily through commerce. As such, the dissemination and use of knowledge is highly sensitive to cost, marketing, and access; yet it may be even more sensitive to less tangible traits of values and culture—human capital in the broadest sense.

Too often, when institutions serve as agents of technology transfer, they focus only on the actual product, process, or service innovation being installed. Does it work when applied? How much does it cost? What problems might it solve? What training is required to use it? Demand questions are harder to assess, though arguably they are even more fundamental. Will people want to use it? Does the population possess high "tinkering skills" as well as the more easily measured skills of literacy and numeracy? Are there potential first adapters at the innovation threshold who will be willing (and able) to stretch to do something new? Indeed, do the innovations create a gleam in the eye that will help to shape the career track of young people, encourage productive trial and error on the shop floor, and inspire weekend R&D at the most informal household level?

Like the creation of new knowledge, the costs of innovation adoption at the base of the population go far beyond the cost of new innovations. In the agricultural sector, a great deal is at stake when a farmer is asked to abandon proven methods that have been used for generations. That is why demonstration projects are necessary. To put new technology to use, local research and experimentation must be adapted to microclimates, soils, water conditions, and pests. That has more to do with the adaptive skills of farmers than the technical training of agricultural extension agents. In Brazil today, basic literacy and numeracy are widespread in all but the most remote rural areas thanks to decades of effort to provide primary schooling. However, schools have been less successful in universalizing critical reasoning, flexible thinking, and day-to-day application of scientific method. In this regard, despite its high level of functional literacy, Brazil's capacity for local R&D may still be at a significant disadvantage.

The same principles apply to industrial technologies. They too must be adapted to local conditions, including the availability of raw materials, unique characteristics of the productive environment, and idiosyncrasies in sources of energy, climate, and the labor force. In some countries, such as Japan, local prefectures often set up their own R&D labs to help firms adapt industrial technology to local conditions. In other countries similar functions are carried out by productivity centers, university technology outreach centers, and private engineering and consulting firms. Most important, workers are expected not only to understand the new technologies, but to be part of the process.

In services, technology is generally disseminated through direct interaction with users. The use of new technology usually requires basic literacy as a prerequisite to specialized training. Beyond literacy and specific skills, prospective users may then require access to complementary inputs and supporting industries that are not otherwise available.

Technology is typically embedded in something that is sold—whether as new equipment, inputs, or training—and financing is often required to purchase it. At the firm level, this may mean financing to buy a license, build a plant, or expand an enterprise. The government can help to do that in the way that it shapes the nation's investment climate (its enabling environment). Government policy can also encourage broader-scale use and adoption of innovation—for example, policies to encourage Internet and computer applications at the school, small business, and even household levels.

Brazil's National Innovation System: Instruments, Institutions, and Human Capabilities

Table 3.1 provides a schematic snapshot of Brazil's innovation system, tying together what has been discussed thus far and providing a blueprint for the analysis to follow. Each element will be taken up and discussed more fully in succeeding chapters, and cross-references are indicated in each cell in italics.

Table 3.1. The National Innovation System of Brazil: Instruments, Institutions, and Human Capabilities

Types of innovation	Policies and instruments	Institutions	Human capabilities
Creating and commercializing knowledge	• Public spending on R&D, including national mission programs, competitive R&D grants, and peer review • Public policies for R&D, including matching R&D grants, tax incentives for R&D, and the IPR regime • Public policies for the commercialization of publicly financed knowledge, including new innovation law and partnerships with the private sector as a condition of research grants • National research education networks *Discussed in chapter 4*	• Public R&D in government labs and universities • Private R&D in private labs, firms, and private universities • Informal innovation in private firms • Specialized government agencies supporting creation and commercialization of knowledge (e.g., CNPq, FINEP, BNDES) • Specialized NGO innovation institutions (e.g., FAPESP) • Intellectual property institutions (e.g., INPI) • Technology transfer officers in public R&D labs and universities • Scientific/industrial parks • Business incubators • Early stage finance and venture capital *Discussed in chapters 4, 5, and 6*	• High-level productive capacity among scientists, engineers, and technicians • Capacity to educate new generations in cutting-edge research • Business leaders and managers who understand high-level science and technology • A culture of techno-entrepreneurship *Discussed in chapter 6*
Acquiring foreign knowledge and technology	• Openness to outside, including trade, foreign direct investment, and technology import policy • Foreign acquisition through education, business travel, trade shows, publications and databases, and Internet access • Incentives to bring the "Brazilian Diaspora" home • Setting up R&D antennae abroad *Discussed in chapter 4*	• Firms willing to purchase embedded technology • University exchanges and foreign collaboration • Federal, state, and municipal sponsorship, purchase, adoption • International NGOs transferring technology to civil society • Individual consumers purchasing technology • Internet-based technology acquisition *Discussed in chapters 4 and 5*	• Knowing "what to look for," including: • Global scanning • Technology assessment • Technology negotiation • Adaptation to domestic conditions • Cost-benefit analysis of technology acquisition *Discussed in chapter 6*

Diffusing and using knowledge that exists within-country	• Public polices setting up technological information and technology extension services • Policies on standards • Policies on intellectual property rights • Strategies to broaden Internet access *Discussed in chapter 4*	• Technical information services • Extension services in agriculture, industry, and services • Productivity organizations • Technology support institutions and programs such as SEBRAE • Metrology, standards, and quality-control systems • Industrial clusters *Discussed in chapters 4 and 5*	• Basic literacy and numeracy • Computer literacy • Communication skills • Updated vocational skills • A "culture of curiosity"; respect for "outside-the-box" thinking • Pervasive understanding of the scientific method *Discussed in chapter 6*
Enabling environment (i.e., the investment climate)	• Competition and trade policy • Effective regulatory policy • Support for entrepreneurship • Good rule of law • Good macrostability *Discussed in chapter 2, though all chapters are relevant*	• Efficient financial system • Flexible labor markets • Fair courts and justice system • Effective governance • Effective formal education institutions and lifelong learning system *All chapters as relevant*	• Capabilities to ensure macroeconomic stability, rule of law, security, and efficient capital and labor markets • Basic citizenship skills • Education and skills required to compete in an increasingly demanding global economy *Discussed in chapter 6*

Source: Authors.
Note: BNDES = National Bank for Economic and Social Development; CNPq = Conselho Nacional de Desenvolvimento Científico e Tecnológico; FAPESP = Fundação de Amparo à Pesquisa do Estado de São Paulo (São Paulo State Research Foundation); FINEP = Financiadora de Estudos e Projetos (Financier of Studies and Projects); INPI = National Institute of Intellectual Property; NGO = nongovernmental organization; SEBRAE = Brazilian Service for Assistance to Small Business.

CHAPTER 4

Assessing Innovation at the National Level

At the macro level, how well is Brazil doing with the three kinds of innovation activities? Creating new conceptual knowledge through research and development in Brazil has been relatively brisk—as measured, for example, by publications in refereed scientific journals. However, R&D has been much less successful in energizing production of technological innovations—for example, patents that can be commercialized.

This chapter looks at Brazil's national innovation system from the perspective of the three types of innovation, beginning with a macro-level comparison of Brazil's efforts to create and commercialize knowledge with efforts by the BRICs (Brazil, the Russian Federation, India, and China)[1] and the BRICKMs (Brazil, Russia, India, China, the Republic of Korea, and Mexico). A closer look is then taken at how Brazil and Russia have responded to the challenges of enhancing national innovation capacity. In assessing the acquisition of foreign technology, Brazil is compared with the other members of the BRICKM grouping. Some macro evidence is then provided on dissemination and use of technology in the manufacturing sector, as revealed by sectoral and firm productivity data.

The chapter concludes with a broad overview of Brazil's national system of innovation, primarily as conceived by government policy makers, including its history and intellectual roots. Brazil started an innovation system earlier than most other then-developing countries, yet its conception of innovation was (and still is) rooted quite narrowly—with a heavy emphasis on creating new knowledge rather than acquiring and adapting what already existed. Despite its notable islands of R&D excellence, Brazil is generally underperforming in innovation. Trade policies that protected domestic producers from foreign competition have exacerbated the deficit, undercutting the need for the private sector to invest in R&D or its commercial applications. These two weaknesses lie at the heart of Brazil's lackluster economic growth in recent years.

Carl Dahlman was a key contributor to this chapter.

Comparative Assessment of R&D—Inputs and Outputs

In recent years, Brazil has experienced considerable scientific and technological success. The number of full-time researchers increased more than sevenfold from 21,500 in 1993 to 158,000 in 2004.[2] Brazil's share of global scientific publications nearly tripled from 0.64 percent in 1990 to 1.73 percent in 2004, with particularly impressive accomplishments in the agricultural sciences (3.08 percent), physics (2.48 percent), pharmacology (2.41 percent), microbiology (2.33 percent), and aeronautics and space sciences (2.11 percent).[3] Illustrating the broad diversity of accomplishment, Brazilian contributions have ranged from cracking the genetic code of the *Xylella fastidiosa* (a bacterium that attacks orange trees and vines) to world-class technology programs in aeronautics (Embraer), satellites (China-Brazil Earth Resources Satellite Program), biotechnology (Genoma), agriculture (Embrapa), and deepwater oil exploration. Petrobrás, to take an example, held 160 patents in the United States in 2005,[4] and in 1996 Embrapa accounted for half of all agricultural research spending in Latin America (Beintema et al. 2001).[5]

Brazil's technological performance stands out in Latin America, but its performance is poor compared with OECD economies.[6] Using worldwide measures of technology performance, Brazil is in an intermediate position. On the Networked Readiness Index (NRI), which measures a nation's readiness to participate in and benefit from information and communication technology developments, Brazil ranked 46th among 104 countries in 2004–05—ahead of Indonesia (51st) and Mexico (60th) but behind Singapore (1st), Korea (24th), Chile (35th), India (39th), and China (41st) (Dutta and Lopez-Claros 2005).

Brazil's performance appears to be related less to input shortage and more to the character of its R&D expenditure. As figure 4.1 shows, R&D investment as a share of GDP in 2004 was relatively high for Brazil's development level—greater than Italy's, Spain's, or Portugal's. However, despite the relatively high total, the distribution of investors was heavily weighted toward the public sector—55 percent, compared with 30 percent in the United States.

Although Brazil's expenditure on R&D as a share of GDP is above the average for its level of per capita income, it is low compared with China and India, two of the most important BRIC comparators and countries with much lower per capita incomes.

In India, the share of R&D to GDP has increased from 0.8 percent of GDP to 1.1 percent since 2005. This is due to a significant increase in private investment, led primarily by a dramatic rise in R&D centers of multinational corporations. The jump in investment is not confined to foreign companies, however. Indian firms gradually have increased their R&D-to-sales ratios since liberalization of the country's trade regime in the early 1990s. In the past two years, they have sharply increased R&D expenditures, having witnessed the benefits multinational companies have reaped from R&D in the face of stiff competition. This has been particularly true for Indian pharmaceutical companies since India extended protection to product patents (see World Bank 2007).

Figure 4.1. GDP Per Capita versus R&D Expenditure as a Share of GDP for LAC and Select Countries, 2004

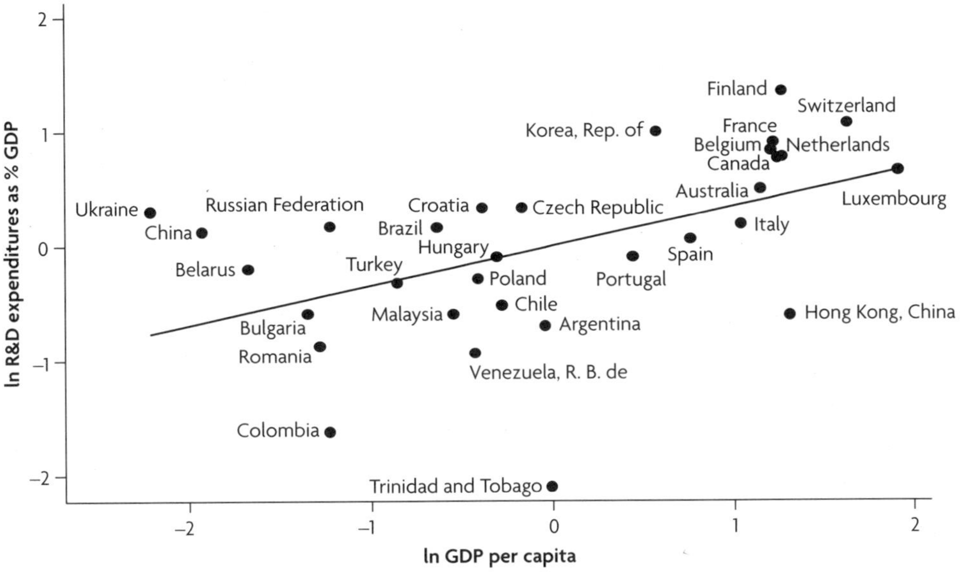

Source: Based on data from the World Development Indicators (WDI) Database.

China increased its share of R&D spending from 0.8 percent of GDP in 1995 to 1.1 percent in 2002. Spending was ramped up further in 2003, and by the end of 2006, it had reached 1.6 percent of GDP. In terms of PPP, China is the world's second largest investor in R&D, trailing only the United States. Although China's R&D spending is still inefficient, 65 percent is already being carried out by enterprises, and a major effort has been launched to improve efficiency. According to China's 15-year science and technology plan, R&D will climb to 2 percent of GDP by 2010 and will reach 2.5 percent, the average level of advanced countries, by 2020.

If it hopes to keep up, Brazil must increase both public and private investment in R&D. Efficiency of public spending needs to be improved at the same time, especially in the production of practical technological innovations.

Brazil has a large number of researchers, partly because it has a large population. However, as shown in figure 4.2, the number is also large in a relative sense; so it is important to understand how well researchers are being used.

One measure of research effectiveness is the ratio between R&D expenditure and patents obtained in the United States. In this area, too, Brazil is lagging (table 4.1).

Brazil's modest R&D performance may be partly explained by the relatively large share of R&D that occurs within universities. International evidence suggests an inverse relationship between the level and effectiveness of R&D when R&D is not linked to strong incentive regimes. This applies to both universities and public institutes. In Brazil, the incentive regime for research is misaligned. First, expenditure is not geared toward cost-effective, output-oriented

Figure 4.2. R&D Expenditures as a Share of GDP versus Researchers in R&D per Million People in LAC and Select Countries, 2004

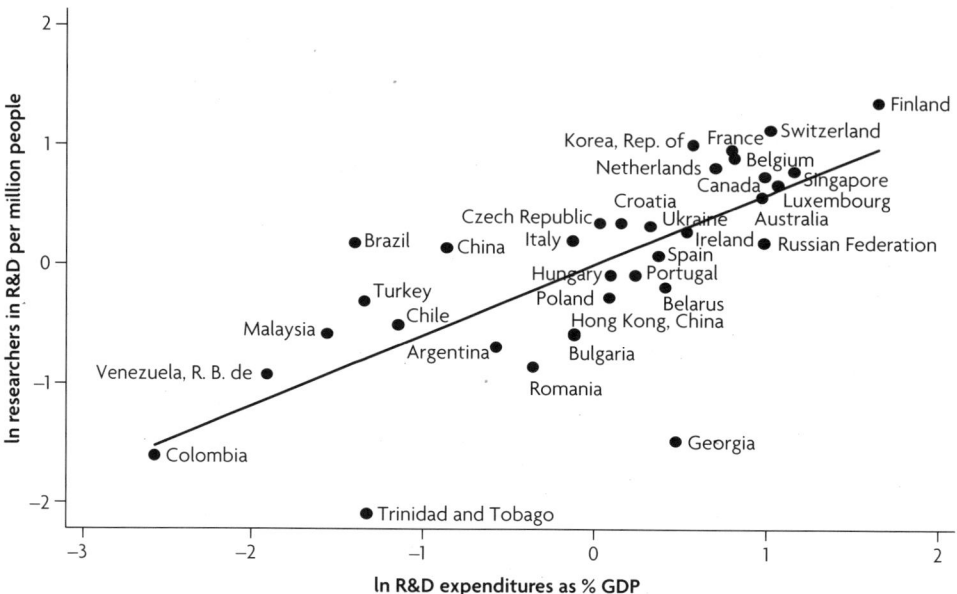

Source: Based on data from the WDI Database.

Table 4.1. R&D Expenditures and Efficiency in Brazil and Select Countries, 2003

Country	R&D expenditures		R&D effectiveness
	% GDP	Per researcher (constant 2000 US$)	Patents per million (US$) of R&D expenditure
Brazil	0.98	76,967	0.51
Canada	2.06	246,184	0.26
Chile	0.70	133,068	—
China	1.22	4,508	2.30
India	0.85	—	—
Korea, Rep. of	2.53	69,431	5.32
Mexico	0.40	37,039	0.23
Singapore	2.15	133,515	0.26
Spain	1.03	33,577	0.43
United States	2.65	297,211	0.68

Source: Based on data from the WDI Database.
Note: — = not available.

research. Second, there is little expectation that publicly generated knowledge will be transferred for commercial application to realize productivity gains. Because most research budgets are funded through earmarked government resources, universities and public research centers have few incentives to orient themselves toward private sector needs. It is no accident that Embrapa—a notable exception in the high effectiveness of its R&D—also obtains a significant share of its research budget through competitive bidding.[7]

Conditions governing intellectual property rights law remain inadequate despite recent improvements, especially in the licensing of processes (which has been simplified since 1993). The National Institute for Intellectual Property (*Instituto Nacional de Propriedade Intelectual*, INPI) still lacks appropriate human and financial resources. This deficiency leads to lengthy licensing processes, thereby reducing the ability of private entities to appropriate investment. Tax breaks and incentives for R&D are similar to those of developed countries—for example, accelerated R&D depreciation allowances and carry-forward provisions. However, innovation policy works in the opposite direction. Considering the typical public-goods problem associated with the inability of private investors to appropriate R&D expenditures, the lack of a properly designed innovation policy helps to explain the discrepancy between social returns on R&D and low private expenditures.[8]

Table 4.2 compares key R&D input and output indicators in the BRICKM group of countries. Brazil compares favorably with Mexico but trails far behind other countries on most indicators (except those scaled by population, such as China and India). Brazil does better than China or Korea or Mexico in amount spent per scientific and technical journal article, but it lags behind all countries except China and Russia in the amount spent per patent granted in the United States. R&D in Brazil has been effective in generating conceptual knowledge but relatively ineffective in generating technological innovations, as measured by the number of patents granted annually. This gap highlights a general disconnect between universities and firms.

How Brazil and Russia Face Innovation and Performance Challenges[9]

Brazil suffers from innovation and performance challenges similar to Russia's, but to a lesser degree. Reviewing the Russian experience, therefore, has relevant policy implications for Brazil given the similarities between the two countries (see table 4.3).

The Soviet Union was, of course, a superpower, based primarily on its military applications of the country's strong scientific and technological capability. However, the inability to compete economically with the United States became a fundamental reason for its break-up after 1991. The transition since 1991 has been extremely painful. Russia's GDP fell significantly between 1990 and 1996. A recovery began in 1997 but a crash followed in 1998 in the aftermath of the 1997 Asian financial crisis. In 1998 the government defaulted on its debt and devalued the currency.

Table 4.2. R&D Inputs and Outputs for the BRICKM Country Group

Indicator	Brazil	Russia	India	China	Korea	Mexico
Researchers in R&D, 2003	59,838	477,647	117,528	926,252	151,254	26,800
R&D researchers per million population, 2004	344	3.319	119	708	3,187	268
Spending on R&D (US$ billions), 2004	5.9	6.8	5.9	27.8	17.9	2.7
Spending on R&D (% of GDP), 2004	0.98	1.17	0.85	1.44	2.65	0.43
Scientific and technical journal articles, 2003	8,684	15,782	12,774	29,186	13,746	3,747
R&D spending (US$ thousands) per scientific and technical article[a]	682	431	460	953	1,332	722
Scientific and technical journal articles per million population, 2003	47.9	109.1	12.0	22.7	287.5	37.1
Patents granted by U.S. Patent Office, 2004	161	173	376	597	4,671	102
R&D spending (US$ millions) per patent granted[a]	36.6	39.3	15.6	46.6	3.8	26.9
Patent applications granted by U.S. Patent Office per million population, 2004	0.90	1.21	0.35	0.46	97.03	0.98

Source: Compiled from data in KAM (2006) and World Bank (2006d).

a. Calculated by dividing estimated R&D spending in 2004 by the number of articles or patents.

With the weak ruble complemented by rising commodity exports, the economy began to grow in 1999 and has continued to do so. The average growth rate in 1999–2007 was 6.7 percent, led primarily by the rapidly expanding petroleum sector. Oil, natural gas, metals, and timber account for over 80 percent of Russian exports. The major noncommodity exports are chemicals and military equipment. Thanks to its strong commodity-based exports, the country has been running a large trade surplus. Russia created a stabilization fund in January 2004 to reduce the rapid appreciation of the ruble and has been paying off its foreign debt from this fund. Nevertheless the strong trade surplus from persistently high oil prices continues to push the currency upward.

In 1990, before the breakup of the Soviet Union, what is now the Russian Federation spent 2.03 percent of GDP on R&D and had 1.9 million scientists and engineers. With the recession following the transition after 1991, R&D spending plunged by over 80 percent in real terms to 1.06 percent of a much smaller GDP in 1999. The number of scientists and engineers in R&D fell to 872,000 by 1998. Spending on R&D has since increased as a percentage of GDP; total outlays are now slightly higher than Brazil's.

The Russian R&D system before the breakup of the Soviet Union was state centric—R&D was carried out in universities, public research centers, and labs

Table 4.3. Basic Comparisons between Brazil and Russia

	Brazil	Russia
Population (million)	180	143
Gross national income (billion)	662.0	638.1
GNI/capita	3,550	4,460
GNI PPP (billion)	1,534.1	1,522.7
GNI PPP/capita	8,230	10,640
Merchandise exports	118,308	243,569
Manufactured exports (%)	54	19
Manufactured exports (billion)	63,886	43,278
Tertiary enrollment coverage (%)	22	68
R&D/GDP	0.98	1.17
Researchers in R&D/million population	344	3,319
Scientific and technical papers in 2003	8,684	15,782
Patents granted in U.S./million population	0.75	1.34
High-technology exports/manufactured exports	13	8
WEF Global Competitiveness Index	66	62
Basic requirements	87	66
Efficiency enhancers	57	60
Innovation factors	38	71

Sources: Based on data from the WEF Global Competitivenes Report, WDI Database, and World Bank internal data.
Note: Figures are for 2005 unless otherwise noted. Monetary amounts are in U.S. dollars.

in industrial ministries. University research focused on basic science. Public research centers and ministry labs were oriented toward military applications and were isolated from commerce. Much of this effort was carried out in over 50 science cities walled off from the country's daily life. Since the transition, some effort has been made to reorient research toward commercial needs, but in 1999 the government was still funding more than 55 percent of R&D (as does Brazil currently), little of which targeted the needs of the productive sector. Firms needing technology purchased it from abroad. The public R&D sector found more demand from foreigners than from local companies.

The Russian industrial plant, except in a few sectors (particularly oil and gas) is outmoded if not obsolete. Few manufactured products are globally competitive except for military items (although their pricing is probably not on full commercial terms). Increasing revenues from natural resources and the continued appreciation of the ruble have allowed Russia to import much of the food and manufactured products it needs. In effect, the Russian economy is rapidly deindustrializing (if the oil and gas sectors are excluded from industry) and becoming increasingly dependent on natural resource and arms sales. While this situation is more extreme than in Brazil, the similarities (substitute agricultural commodity exports for oil and gas exports, and Embraer airplane sales for MIG fighter and other armament sales) carry an important lesson for Brazil about the underlying risk of its current commodity export boom.

The main lesson Brazil should draw is that public R&D must be better managed and oriented toward broader economic ends. Russia's tremendous scientific and technological capability and high human capital did not serve it well because these resources were not oriented toward productive needs. Combined with a poor economic and institutional regime, this led to the collapse of the Soviet Union. Even today there is a major disconnect between a much smaller R&D base and the economic requirements of the Russian Federation. The government is trying to reorient its R&D capability toward global economic competitiveness, and it is installing support infrastructure to commercialize the knowledge that is produced. This includes science and technology parks, business incubators, venture capital for high-tech start-ups, and stronger intellectual property and licensing procedures. However, despite the great progress in improving the macroeconomic situation, Russia still suffers from a poor business environment and porous rule of law. As a result, the capacity of the domestic science and technology system to create wealth has not been harnessed, and Russia earns poorer marks on innovation capability than Brazil in World Economic Forum (WEF) rankings (see table 4.3).[10]

The risk Russia illustrates for Brazil is that a continued agricultural commodity boom may divert attention away from retooling the manufacturing and service sectors for greater competitiveness, resulting in an overspecialization in natural resources. This has implications for Brazil's exchange rate strategy and highlights the need for improvement in its overall business environment and innovation capability.

Figure 4.3 provides a comparative perspective of Brazilian innovation assets compared with Russian innovation assets.

Figure 4.3. Brazil and the Russian Federation Innovation Assets in Comparative Perspective

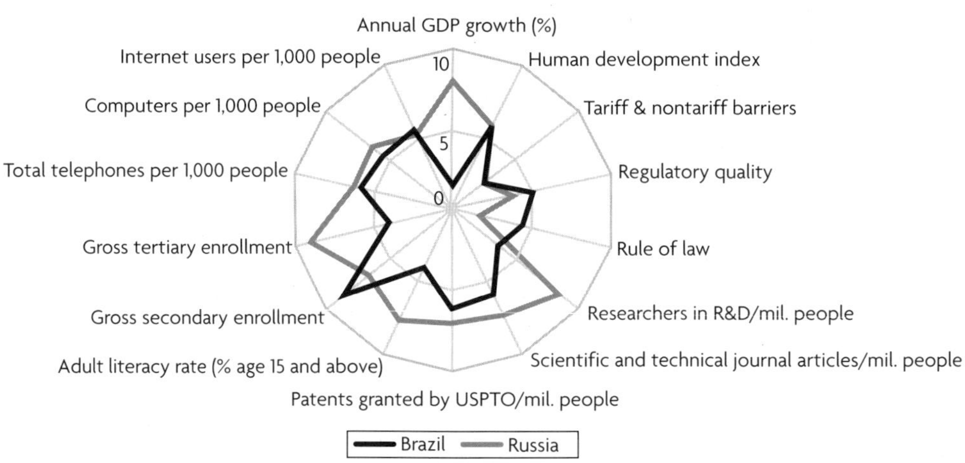

Source: Calculated using the world Bank's online KAM tool.

Acquisition of Foreign Knowledge

Table 4.4 shows data on how foreign knowledge is acquired by the BRICKMs. Along with India, Brazil stands out as making the least use of foreign knowledge through means other than FDI. The most striking element concerns trade. Brazil is among the most closed of the major economies, both in the low share of the economy that is traded and the high degree of protectionism.

The combination of low investment in GDP compounded by low capital goods imports is among the most significant constraints to acquiring global knowledge. Brazil acquires less foreign innovation through capital goods than do its peers (figure 4.4), so technological change is less absorbed by manufacturing firms. Increased integration could potentially lead to higher imports of these goods, helping to both develop innovations at the firm level and increase productivity.

The government has undertaken licensing-agreement and capital-good-acquisition initiatives to expand technology absorption. The process of deregulating technology transfer started in 1991, with further steps taken in 1993. INPI registration time for contracts was shortened, and several administrative procedures were waived. This partly explains the boom in royalty payments in the late 1990s, which increased from 1 percent of GDP in 1990 to 8 percent in 1995 to 24 percent in 1999, before retreating to 18 percent in 2005 (World Bank 2006). Unnecessary requirements still stymie the process, and further simplification is needed.[11] Import tariffs and the tax on manufacturing goods (IPI), which applied to capital goods imported by exporters, were reduced after 2000 and eliminated in June 2005 (Decree No. 5,468). Limited

Table 4.4. Comparative Data on Acquiring Foreign Knowledge (BRICKMs)

	Brazil	Russia	India	China	Korea	Mexico
Trade as share of GDP (2004)	31.40	57.30	41.60	65.30	83.80	62.00
Tariff and nontariff barriers (2006)	3.50	3.50	5.00	3.00	3.50	2.50
Gross foreign investment as share of GDP (avg. 2000–04)	3.72	1.36	0.68	3.89	1.04	2.80
Royalty and license-fee payments (US$ millions, 2004)	1,196.9	1,095.4	420.8	3,548.1	4,450.3	805.0
Royalty and license-fee payments/ million pop. (2004)	6.70	7.66	0.40	2.75	92.52	7.76
Manufacture trade as share of GDP (2004)	16.38	17.85	15.29	50.35	55.30	47.54
High-technology exports as share of manufacture trade (2003)	11.96	18.86	4.75	27.10	32.15	21.34

Source: WDI 2006 and KAM 2006.

Figure 4.4. Total Imports versus Imports of Capital Goods in LAC and Select Countries, 2004

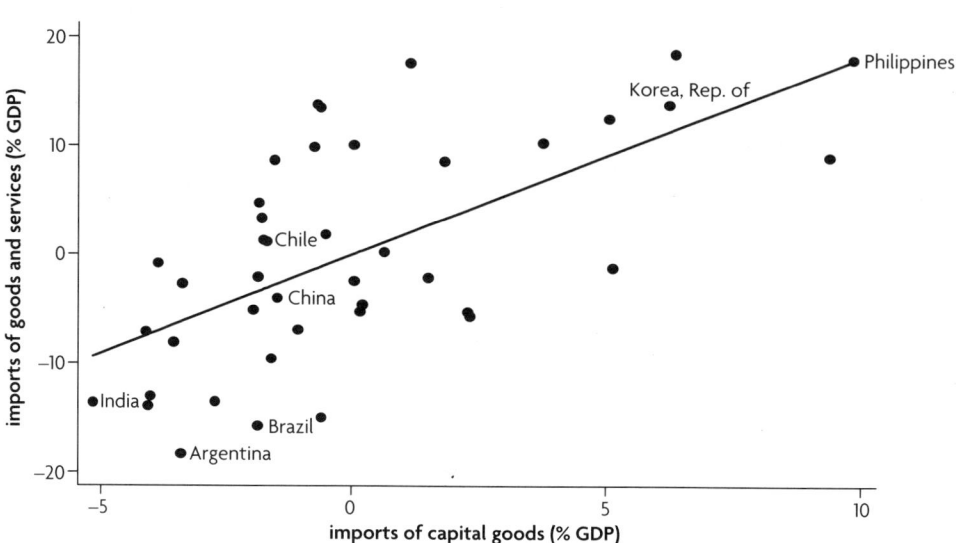

Source: Based on data from the WDI Database.

Box 4.1. The Statute of Microenterprises and Small Enterprises

Microenterprises and small enterprises account for about 90 percent of firms in Brazil. Reflecting their important role, the Statute of MSEs (Complementary Law No. 123) was approved in December 2006, following a long period of discussions between the public and private sectors. A key provision in the law mandates that all levels of government, their respective funding agencies, and the centers for innovation develop programs targeted to assist MSEs. These programs should receive at least 20 percent of agency funding for innovation, which must be documented in each institution's annual report. In addition, taxes on capital-goods purchases by MSEs were eliminated to spur technology absorption within the sector.

access to credit, especially for micro- and small enterprises (MSEs), constrains the import of capital goods through differential tax treatment of capital-goods purchases. An initiative that should help, the Statute of Microenterprises and Small Enterprises (box 4.1), was approved at the end of 2006.

The Dissemination and Use of Knowledge

It is difficult to make a cross-country macrocomparison of knowledge dissemination and use. However, some patterns appear when looking at firm-level data recently compiled through the World Bank Investment Climate Survey.

Table 4.5. Productivity Dispersion within Brazil's Industrial Sectors
value added per worker

Sector	Max/Min	Adj Max as % of Max	Adj Max/Mean
Food & Beverage	12,900.07	57.22	9.42
Textile	1,169.01	67.31	5.99
Apparel	79,103.56	31.60	9.14
Leather & Footwear	65,897.30	73.33	4.81
Chemicals	9,879.34	61.91	7.83
Machinery & Equipment	315,929.99	37.98	33.83
Electronics	6,658.67	52.03	10.00
Auto Parts	689.60	64.88	4.17
Furniture	26,916.31	35.06	7.88
Average	57,682.65	53.48	10.34

Source: Computed from ICS.
Note: The top and bottom 1 percent for the sample were discarded to eliminate false readings from data errors.

Table 4.5 shows the high dispersion of value added per worker across nine representative industrial sectors in Brazil. Particularly striking is how large the difference is between the most- and least-efficient firms, with the greatest disparity—300,000 times more value added per worker—occurring in the machinery and equipment sector. The average for all nine sectors is an amazing 57,000 times more value added per worker. To obtain a conservative measure less influenced by outliers, the maximum was adjusted by taking as the maximum the value of the dense part of the distribution. The adjusted maximum averaged 53 percent of the distance to the recorded maximum. Even with these conservative adjustments, it appears that if average productivity could be raised to the adjusted maximum level, it would increase by a factor of 10.[12]

This analysis suggests just how much national output could be raised—at least in principle—if all Brazilian firms adopted existing technology. Obviously, moving to higher-productivity technologies is not costless. Firms that currently use such technologies are likely to be much larger, use other modern equipment, employ more up-to-date management practices, use better inputs, and have better educated and more skilled workers compared with firms that do not use them. The latter, more typical firms are operating far behind their more efficient counterparts. Far more must and can be done to disseminate and effectively employ existing knowledge across the board.

Data from the Investment Climate Survey (ICS) make it possible to analyze the dispersion of labor productivity across industrial sectors in other countries. It is surprising that the productivity dispersions are on average, twice as large in Brazil as in India, considering that dispersions in the latter already exceed those in most of the countries to which it has been compared.[13] If average productivity could be raised to the maximum level in India, using a similar methodology, it would rise only by a factor of 5 compared to a factor of 10 for Brazil.

Countries develop specialized mechanisms and institutions to disseminate knowledge and help firms use it effectively. Brazil has been successful in creating and disseminating agricultural technology thorough Embrapa and various specialized, state-level agricultural extension institutions. Brazil also has some specialized institutions that seek to disseminate technology information and training in industry, such as *Serviço Nacional de Aprendizagem Industrial* (SENAI) and other industry associations. It is unclear why Brazil's productivity differentials are so high, and this issue needs more detailed examination.

The recently passed Statute of Microenterprises and Small Enterprises should help MSEs purchase capital goods and target R&D efforts to make knowledge more widely available and usable. However, it is too early to tell how well the initiative is working.

The National System of Innovation as Conceived by Government

Brazil—more than most middle-income countries—has a long tradition of backing R&D.[14] The effort began in the 1950s with limited resources and through indirect means such as investments in public infrastructure (research centers, provision of technical assistance, and metrology services), human resources formation, and other externalities. Significant resources were directed to these areas in the 1970s and early 1980s, led by the military government's desire to achieve some domestic technological capability. By the mid- to late 1980s, a relatively well-structured science and technology (S&T) base was in place, and the results—in terms of more graduate programs, research groups, and scientific publications—began to appear.

The 1980s witnessed the first significant attempts to support company-based R&D. A number of direct instruments were added, including fiscal incentives, credit channeled to firms by the Financier of Studies and Projects (*Financiadora de Estudos e Projetos*, FINEP), and the procurement of targeted goods and services by government-controlled enterprises such as state oil company Petrobrás. Responding to the limited impact these measures achieved in spurring companies' R&D, the government enacted comprehensive legislation to promote market-oriented innovation. The *Programas de Desenvolvimento Tecnológico Industrial and Agropecuário*—PDTI/PDTA (Law 8661/93) provided tax breaks and other incentives for competitive industrial and agricultural research by public and private firms; applications for firm-specific multiyear plans for technology development were filtered through the Ministry of Science and Technology (MCT). The fiscal crisis in the latter 1990s and the need to build a primary budget surplus led the government to cut support for R&D (Law No. 9532/1997), mostly by reducing public infrastructure investment and paring already meager fiscal incentives.[15]

MCT and the National Council on Science and Technology (CCT) have defined S&T strategies and coordinating intergovernmental initiatives since 1995.[16] Two strong federal institutions—the National Council for Scientific and Technological Development (*Conselho Nacional de Desenvolvimento Científico e Tecnológico*, CNPq) and FINEP—were established to promote basic research. They offer graduate and postgraduate programs and finance technological investments by the private sector.[17] CNPq directly manages several research institutes, including the well-regarded Brazilian Center for Physics Research

(*Centro Brasileiro de Pesquisas Físicas*, CBPF) and the National Institute of Basic and Applied Mathematics (*Instituto Nacional de Matemática Pura e Aplicada*, IMPA). Successful research centers are attached to other ministries as well—for example, the Brazilian Agricultural Research Corporation (*Empresa Brasileira de Pesquisa Agropecuária*, Embrapa, box 4.2) and the National Institute for Metrology (*Instituto Nacional de Metrologia*, Inmetro) are maintained by the Ministry of Agriculture and Livestock and the Ministry of Development, Industry, and Trade, respectively.[18] State-owned enterprises also run their own research centers—for example, Petrobrás's Cenpes—while public-private institutions such as SENAI support technology centers.

In a decentralized federation such as Brazil, individual states play an important role in financing R&D, and they have full autonomy in setting their S&T policies. Several have their own support agencies, as well as higher education and research institutions. Estimates from MCT show that the states accounted for 30 percent of government spending on S&T in 2004. São Paulo State has the largest state-level R&D support system, and it is also the largest recipient of federal funds. About two-thirds of R&D public funding in São Paulo state—around 1.1 percent of the state's GDP—comes from state sources, including funding for three state universities, 19 research institutions, and FAPESP, the state's S&T support agency (FAPESP 2004). The strong support by the state government makes São Paulo the second-largest investor in R&D in Latin America, ahead of Mexico and Argentina. Other states active in this area include Rio de Janeiro, Minas Gerais, and Rio Grande do Sul, albeit with much smaller budgets (Cruz and de Mello 2006).[19] Because of the importance of states in Brazil's innovation system, the next step is to apply the conceptual framework used in this study to the analysis of knowledge and innovation for competitiveness at the state level.

Box 4.2. Embrapa

Embrapa was created in 1973 to "develop solutions for sustainable development in Brazil's rural areas, focusing on agribusiness through the creation, adaptation, and transfer of knowledge and technologies to benefit society." It has 37 research centers and 2,221 researchers (53 percent holding PhDs). Most research centers carry out commodity-specific research, while others are involved in thematic research (the environment, genetics, and biotechnology) or regional issues. The corporation also has two laboratories operating overseas (one in France and one in the United States). Embrapa coordinates the National System of Agricultural R&D, including federal and subnational R&D institutions, universities, and businesses. Along with subnational agricultural R&D institutions, Embrapa has helped Brazil become one of the world's largest agricultural producers and a competitive, low-cost exporter of agricultural commodities.

Source: Based on Cruz and de Mello (2006).

The structure of the Brazilian National Innovation System is complex, involving the Ministries of Science and Technology, Education, Health, Agriculture, Development and Foreign Trade, Defense, and others (figure 4.5). At the federal level, CCT an advisory body to the presidency, has a policy coordination role. MCT is the executive body, with assistance from FINEP, CNPq, and the Center for Management and Strategic Studies (CGEE). Industrial policy is formulated by the Ministry of Development, Industry, and Trade (MDIC) through the National Council of Industrial Development (CNDI) and the Brazilian Agency of Industrial Development (ADBI). Coordination between science and technology and industry and commerce is promoted by MCT and MDIC representation in both CCT and CNDI. However, the coordination is not very strong. Moreover, the National Bank for Economic and Social Development (BNDES)—the main financier for development—acts independently. Finally, as this structure makes evident, the Brazilian concept of a national innovation system devotes little explicit attention to acquiring foreign knowledge or to disseminating knowledge, the other two components of the innovation system framework. In fact, it is quite telling that the initial source of funding for the National Fund for Scientific and Technological Development (*Fundo Nacional de Desenvolvimento Científico e Tecnológico*, FNDCT) was a tax on technology imports.

FINEP's creation of 16 S&T sector funds (box 4.3) since 1999 has been the most important attempt to provide a stable, complementary source of public funding for R&D. This was done in the context of increasing fiscal constraints and current expenditures crowding out public investments. FINEP disburses funds through several mechanisms, mainly grants aimed at various economic agents and with multiple objectives. Time periods vary for grants, and the amount disbursed per project is strictly limited.[20] Sector funds are financed by sector-specific contributions and by earmarking royalties and other public revenues.[21] While this has provided a steady source of funding for R&D, earmarking funds to specific sectors works against allocating resources efficiently. Only two of the sector funds, the *Fundo Verde Amarelo* and the *Fundo de Infra-Estructura*, are cross-sectoral. Most funds are used primarily to support universities and research institutes, with little support going to enterprises or to collaborative research with the private sector.

The most complex, controversial, and frequently revised policy instruments are related to private sector tax incentives. Two programs were set up in the early 1990s (though lacking appropriate stimulus mechanisms) to boost technology absorption and diffusion in the manufacturing sector—the Support Program for Technological Industrial Training (*Programa de Apoio à Capacitação Tecnológica da Indústria*, PACTI) and the National Program for Quality and Productivity (*Programa Brasileiro da Qualidade e Produtividade*, PBQP). By the mid-1990s, tax incentives for R&D activities were reinstated for the agricultural sector (PDTA) and the industrial sector (PDTI), and measures to build research infrastructure and train scientific personnel also were promoted. Several technology-oriented measures were put in place with World

Figure 4.5. Brazil's National Innovation System

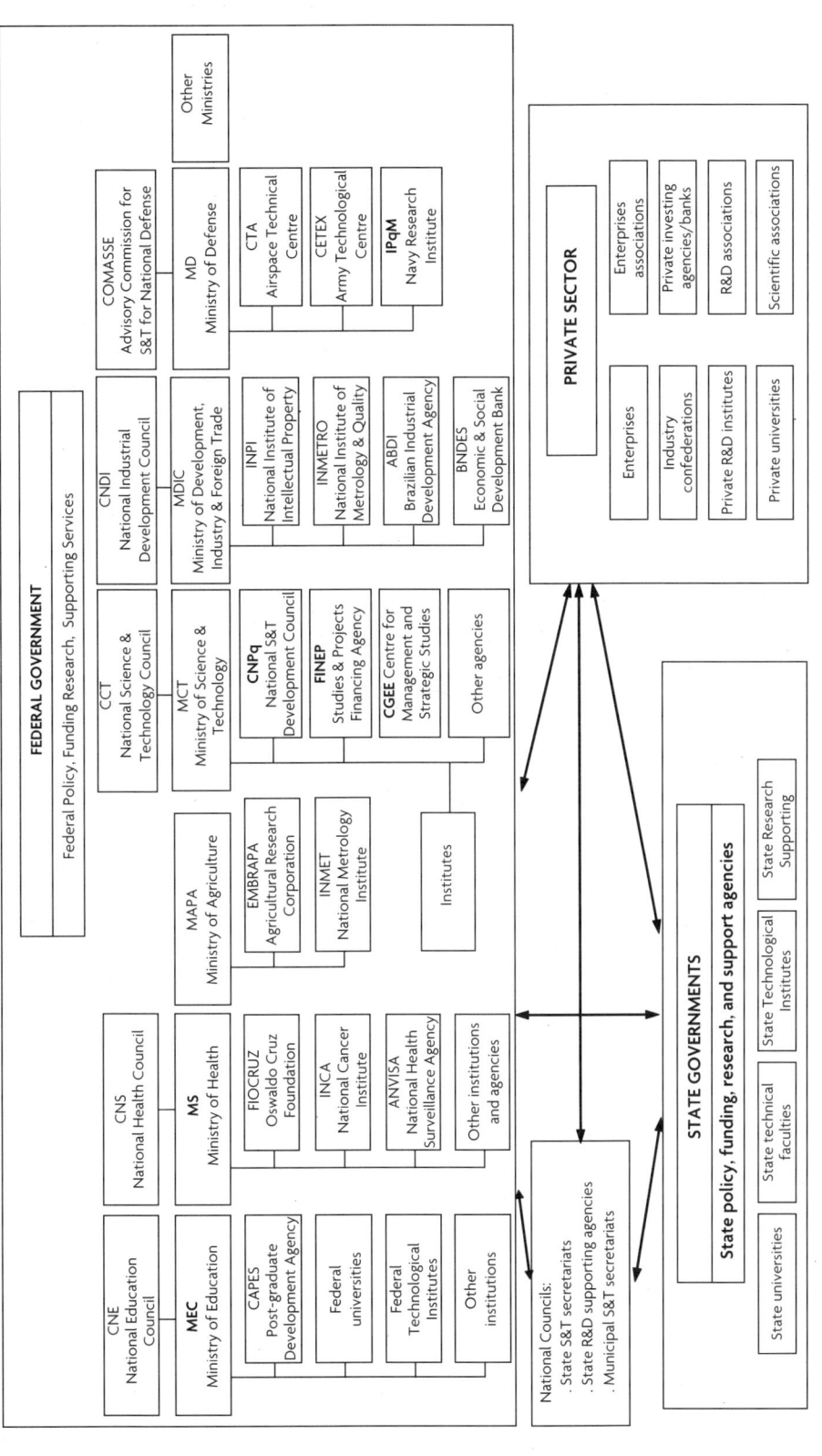

Source: Ministry of Science and Technology.

Box 4.3. The Sector Funds Program

The Sector Funds Program consists of 14 thematic funds (such as petroleum and natural gas, biotechnology, energy, agribusiness, aeronautics, and information technology). Each fund has its own research objectives, ranging from basic research to commercial innovations. Two funds—the University-Industry Collaborations Fund and the Infrastructure Fund—are not related to any particular industrial sector. The latter is designed to improve research facilities, laboratories, and equipment at public research institutions. Sector funds are based on the premise that thematic investment will supply industry's demands for innovation-oriented research. Thirty percent of the funds' resources must be directed to proposals from the North and Northeast regions, which have traditionally relied on federal assistance to establish and sustain research institutions. The rules governing sector fund finance determine that nonprofit universities and research institutes must perform the R&D, while for-profit universities are excluded, and participating businesses must collaborate with a lead nonprofit research institution.

The thematic funds draw their capital from taxation on business profits, royalties, and the use of Brazil's natural resources. For instance, the Petroleum Fund, the first to be regulated in 1999, uses tax revenues from the oil and natural gas industry to finance R&D activities in the industry. Twenty percent of each fund's allocation is pooled in the Infrastructure Fund. All tax revenues are channeled into the MCT's National Fund for S&T Development (*Fundo Nacional de Desenvolvimento da Ciência e Tecnologia*, FNDCT), which has been active since the 1970s.

Operations of the sector funds are administered by FINEP. Independent management committees in charge of fund strategies represent the scientific community, the private sector, and the government. In 2001 a nonprofit organization, the Center for Management and Strategic Studies (CGEE), was hired by the MCT to administer the management committees and provide policy advice. Currently, an umbrella committee, formed by the heads of the management committees, fills that role. Conceptually, sector funds occupy an important niche in R&D sponsorship and industrial development, but other public agencies also support innovation through investment in R&D.

Source: Based on Sá (2005).

Bank support through the Support Program for Scientific and Technological Development (*Programa de Apoio ao Desenvolvimento Científico e Tecnológico*, PADCT), which invested US$470 million in almost 4,500 projects.

In 2006, the revenue foregone through tax incentives for R&D in Brazil was estimated at R$1.6 billion (or 0.1 percent of GDP). Federal laws provide some tax breaks for R&D activities (table 4.6), with most of these incentives targeting the information and communications technology (ICT) industry (Law No. 8,248/1991, altered by Law No. 10,176/2001). Support was subsequently extended to non-ICT firms (Law No. 8,661/1993, amended by Law No. 9,532/1997 and now revoked). Tax breaks are also granted (Law Nos. 8,010/1990 and 8,032/1990) to universities and for purchasing research materials. The national tax code was modified by Law No. 11,196/2005 (MP do Bem), which simplified procedures for firms to claim tax breaks. This measure was welcomed by the private sector, though it is too soon to assess its impact on innovation intensity. As described by Cruz and de Mello (2006), tax benefits include (a) exemption from federal indirect taxes on sales of selected products and purchases of capital goods and intermediate inputs,

Table 4.6. Brazilian R&D Tax Incentives, 2000–06

current R$ million

Law No.	2000	2001	2002	2003	2004	2005*	2006*	Scope
8,010/1990	60.3	118.4	111.9	152.0	155.9	117.8	149.9	Research materials for universities
8,032/1990	10.5	6.3	6.5	8.2	11.4	8.2	11.0	Research materials for universities
8,248/1991 and 10,176/2001	1,203.7	—	732.9	961.7	934.6	1,369.1	1,515.1	R&D in ICT companies
8,661/1993 and 9.532/1997	22.3	22.4	15.2	19.7	37.1	46.1	124.6	R&D in non-ICT companies
8,387/1991	13.4	62.4	77.6	98.1	89.5	96.5	111.0	R&D in ICT companies in Manaus Free Trade Zone
Total	1,310.2	209.5	944.1	1,239.7	1,228.5	1,637.7	1,911.6	

Source: SIAFI, *Sistema Integrado de Administração Financeira do Governo Federal* (Integrated Federal System for Financial Administration).
Note: * = estimates, — = not available.

(b) corporate income tax deductibility for spending on R&D and for payments of royalties for the use of trademarks/patents and technical/scientific assistance, and (c) accelerated depreciation and amortization provisions.

However, only a small fraction of the total public support for R&D through either grants or tax incentives supports work carried out by the productive sectors.

To promote commercialization of knowledge produced by public research institutes and universities, and to encourage greater collaboration between firms, public research centers, and universities, the government passed the Innovation Law in 2004 (box 4.4). While it is a step in the right direction, administrative rules to implement it have not been issued. It has had virtually no impact because firms are reluctant to act, leery of how its provisions will be interpreted by tax authorities.

Conclusion

Although Brazil started to develop a national innovation system quite early, it was narrowly focused on domestic R&D and had a somewhat autarkic focus. Brazil is not getting as much for its R&D efforts as it should, partly because, until recently, the innovation system focused on public R&D labs and universities. With limited exceptions, such as health, agriculture, petroleum, and aeronautics, there were few explicit instructions to the R&D infrastructure to develop practical knowledge, and there were no incentives for the public R&D infrastructure to commercialize the knowledge produced. In addition, the productive sector, operating in a relatively protected economy until the 1990s, had little incentive to undertake R&D to improve its competitiveness

Box 4.4. The Innovation Law

In December 2004, the Brazilian Congress approved the Innovation Law (Law No. 10,973). Although modest in scope and depth, it improves the incentive regime for greater results-oriented public research and more active transfers for the private sector. It was organized around three premises: (a) the development of an environment conducive to strategic partnerships among universities, technology institutes, and the private sector; (b) incentives for S&T institutes to participate in the innovation process; and (c) incentives for innovation within firms. The law authorizes incentives to foster collaboration between public scientific and technological institutions (STIs) and the private sector. It gives STIs more flexibility to negotiate technology licensing agreements and to strike deals with private enterprises for use of public labs. Public researchers will be free to work for other STIs and continue to receive their regular salaries while carrying out joint projects. They also can request special unpaid leave and join a start-up company to further develop their technologies. The law took effect in mid-2005, but administrative regulations have yet to be passed to provide the legal framework to improve the country's capacity to generate and commercialize technology.

and did not receive much support from government to do so. That is changing, but Brazil still has far to go. Policy makers need to think of the innovation system more broadly to include acquiring foreign knowledge and disseminating and using knowledge. The microevidence presented in the next chapter on the importance of these two additional parts of the innovation system should help raise awareness of their need for explicit attention.

CHAPTER 5

Firm-Level Innovation

This chapter examines innovation at the firm level in Brazil using evidence from two databases—first, the PINTEC database developed by the Brazilian Institute of Geography and Statistics (IBGE), and second, the World Bank Investment Climate Survey (ICS) database.[1] The first section compares Brazil's innovation performance with OECD countries. It shows, as expected, that innovation activity is lower in Brazil; but in addition, it shows that the focus of innovation is different, as would be anticipated from the conceptual framework developed in chapter 2.

The second section examines the sources of innovation for Brazilian firms. It finds that the most important source is not R&D but purchases of capital goods, and moreover, that competitive firms and buyers are the most important sources of information about innovation for Brazilian firms.

Based on this microdata, the third section analyzes the relationship among firm-level innovation, productivity, and growth in Brazil. We find that more innovative firms have higher productivity and grow faster than less innovative enterprises. A more formal exploration of the relationship among innovation inputs, outputs, and productivity is done through a three-equation recursive model (presented more fully in appendix A). In brief, we find that innovation (in either of the two ways in which we specify it) positively affects the firm's value added per worker. Both analyses validate the importance of fostering greater innovative effort to improve Brazil's growth.

The fourth section examines firm-level innovation inputs (such as R&D, technology licensing, joint ventures, and worker's training) and outputs (such as new products and new product lines) in Brazil in terms of firm size, sector, and regional distribution. In descriptive analyses, R&D, innovation, and productivity are typically found to increase with the firm's size. The findings show

José Guilherme Reis, Mariam Dayoub, Carl Dahlman, and Paulo Correa were key contributors to this chapter.

that firm size is indeed an important determinant of R&D and innovation for Brazilian manufacturing firms; however, the results for productivity depend on how innovation is defined.

The fifth section summarizes the findings regarding the relationship between various measures of human capital and productivity and various innovation inputs and outputs. This draws attention to the critical role played by education both in acquiring and using existing knowledge, as well as in creating new knowledge. Finally, some conclusions are drawn about firm-level performance insofar as they address the main questions of this report.

Innovation in Brazil: Firm-Level Perspective

In Brazil, firm-level studies recently have been developed, benefiting from the availability of new databases. Indeed, firm-level studies became possible only after 2002, when IBGE released a survey on technological innovation at the firm level (PINTEC), which follows the methodology suggested by the "Oslo Manual," an OECD document establishing guidelines for collecting and interpreting data on industrial innovation. Table 5.1 summarizes the results for data collected in 1998–2000 and 2001–03. The 1998–2000 data is based on 72,000 firms. Only 6.3 percent of firms solely undertook product innovation, 13.9 percent undertook process innovation alone, and 11.3 percent undertook both product and process innovation. Overall, 31.5 percent of firms undertook any of the innovation types cited above. As expected, smaller firms (10–49 workers) undertook fewer product or process innovations than larger firms. The overall percentages did not change much in 2001–03, although the

Table 5.1. Type and Rate of Innovation among Brazilian Firms, 1998–2000 and 2001–03

percent of firms with at least 10 employees

	1998–2000	2001–03
Innovation rate		
Product	6.3	6.4
Process	13.9	12.9
Product and process	11.3	14.0
Any of above type of innovation	31.5	33.3
Innovation rate by size of firms		
Product innovation rate		
Small firms (10–49 employees)	14.1	19.3
Large firms (>500 employees)	59.4	54.3
Process innovation rate		
Small firms (10–49 employees)	21.0	24.8
Large firms (>500 employees)	68.0	64.4

Source: Cruz and De Mello 2006, based on IBGE's Innovation Survey, PINTEC.

percentage of small firms undertaking product or process innovations increased slightly, while that of firms with more than 500 employees decreased slightly.

How do Brazilian firms compare with OECD firms in rate and type of innovation? Based on comparisons with the European Innovation Surveys, Brazil's overall innovation rate is similar to that of Spain but lower than the European average. For example, the innovation rate is 49 percent in Denmark, 51 percent in Holland, 59 percent in Belgium, and 60 percent in Germany.[2] In addition, most Brazilian innovations involve process, not product, innovations; and most of what is reported as innovation in the surveys is actually innovative in terms of being new to the firm rather than new to the market. As noted in chapter 3, this was expected because firms in developing countries like Brazil still stand to benefit very much by tapping preexisting knowledge in the country and abroad.

More detail can be obtained from the database for 1998–2000. The Institute of Applied Economic Research classified firms into three categories to perform a relevant analysis on the data (table 5.2).[3] The categorization sorted firms by their competitive strategy:

- *Group A* firms were those that innovate and differentiate products. These companies carried out innovation new to the market and obtained a price premium equivalent to 30 percent in exported goods when compared to other Brazilian exporters of the same product. R&D, marketing, quality, and brand management were primary emphases.

- *Group B* firms specialized in standard products and adopted a competitive strategy based on cost cutting rather than the value added creation of Group A firms. Group B contains exporting firms not included in Group A and nonexporting firms that are as or more efficient than the exporters. Group B firms seek lower costs and focus on operational manufacturing, management, and control and logistics.

Table 5.2. Basic Characteristics of Brazilian Firms Grouped by Competitive Strategy

	Number and percent of firms	Share of sales (%)	Share of employment (%)	Average number of employees	Average sales (R$ millions)
A. Innovative and product-differentiating firms	1,199 (1.7)	25.9	13.2	545.9	135.5
B. Standard product firms	15,311 (21.3)	62.6	48.7	158.1	25.7
C. Lower productivity firms	55,495 (77.1)	11.5	28.2	34.2	1.3
All firms	72,005 (100)	100	100		

Source: IPEA 2005.
Note: Cells have been left blank where data are not relevant to the analysis.

- *Group* C firms do not differentiate, have lower productivity, and include enterprises that do not fit into Groups A and B. This group comprises non-exporters that are able to perform in less dynamic markets by means of low prices or low salaries.

The innovative and product-differentiating firms compose the smallest group in the PINTEC survey (just 1.7 percent) but account for a quarter of sales and 13.2 percent of employment, making them the largest in average work-force and sales. Firms with standard products are the second most numerous (21.3 percent) but account for 63 percent of sales and 49 percent of jobs and are midrange in average size. Low-productivity firms are the most numerous (77.1 percent) but account for only 28 percent of jobs and 11.5 percent of sales and are the smallest in terms of average employment and sales.

Overall, only 4.1 percent of firms carried out product innovations new to the market, and only 2.8 percent carried out process innovations new to the market (table 5.3). Other innovations were new to the firm but not to the market—in other words, they represented diffusion of technology already available in Brazil. For Group A firms, all product innovations were new to the market. However, it is interesting that 70 percent of them also undertook process innovations, accounting for the highest percentage of process inno-vations new to the market (37.5 percent). This suggests that many product innovations probably required new processes, too. It is also interesting that firms specializing in standard products were associated more frequently with process than product innovations, suggesting that probably they were using existing technology to upgrade their production process to reduce costs.

As noted in chapter 3, much technology is embodied in capital goods. Thus, it is unsurprising that capital goods are the most frequently cited source of innovation by Brazilian firms (table 5. 4). The second most cited source is labor training or the hiring of persons who have the required skills. The third is R&D. Similar results were found by the ICS when around 1,600 firms were asked to identify the most important ways to acquire new technology.

Table 5.3. Type of Innovation by Competitive Strategy of Innovating Firms
percent

Competitive strategy	Product innovation				Process innovation		
	Innovative firms	Subtotal	New for market	New for firm	Subtotal	New for market	New for firm
Group: A: Innovative and differentiating	100.0	100.0	100.0	28.4	70.6	37.5	48.5
Group B: Standard product	44.5	26.3	4.5	23.1	35.6	5.7	31.6
Group C: Less productive	26.4	13.4	1.9	11.7	21.4	1.3	20.4
All	31.5	17.6	4.1	14.4	25.2	2.8	23.3

Source: IPEA 2005.

Table 5.4. Innovation Sources for Brazilian Firms, 1998–2003

percent of firms with at least 10 employees

Source of innovation	1998–2000	2001–03
Acquisition of machinery and equipment	76.6	80.3
Labor training	59.1	54.2
In-house R&D	34.1	20.7

Source: Cruz and De Mello 2006, based on PINTEC.

Of 13 options, the top three selected were (a) acquisition of machinery and equipment (66.4 percent of the firms), (b) in-house development (62.6 percent of the firms), and (c) the hiring of key personnel (45 percent of the firms). Therefore it is important to note that R&D is not the most important source of innovation. This is true even for the most innovative firms in Brazil. The importance of capital goods as a source of innovation at the micro level reinforces the significance of two macro findings reported in chapter 4. These were, first, that low innovation in Brazil is tied to the low investment rate and, second, that Brazil has very low imports of capital goods. The net effect is to deny Brazilian firms access to one of the most important sources of innovation and competitiveness.

Table 5.5 presents the most important source of information for innovating firms according to their type of competitive strategy. It is noteworthy that the most important source of information for all groups of firms is not the in-house research department, but clients and consumers (50 percent of innovating and differentiating firms) or other internal sources such as engineering and maintenance (40–45 percent for the other two groups). The other key sources are suppliers (particularly equipment suppliers for standardized-product and less productive firms) and fairs and expositions (which are rated as being at least as important as, if not more important than, internal R&D for all three categories—especially for standardized-product and less productive firms). The relatively small role played by universities and research institutes is consistent with the findings of the macro assessment made in chapter 4. All of this highlights the importance of promoting competition and technological diffusion to make Brazilian firms more innovative, rather than simply increasing R&D. It is therefore no surprise that Group A firms use information from all the various sources much better than do the other two firm groups.

A final policy-relevant insight from the microanalysis of innovation in Brazil is presented in table 5.6, which lists the main obstacles to innovation as reported by firms. As expected, the most important obstacles are costs, risks, and scarcity of financing. However, it is noteworthy that shortage of skilled workers was reported by almost 50 percent of firms, and that lack of information and difficulty in adopting international standards was reported by a quarter to a third of firms, with the latter reason rising in importance. This highlights the fact that financial constraints are not the sole bottleneck, and policies to foster greater innovation in Brazil must also focus on access to skilled human capital and technological information to be successful.

Table 5.5. Main Source of Innovation Information for Innovating Firms by Type of Competitive Strategy

percent

	Type A (innovating and product differentiating)	Type B (specializing in standard products)	Type C (less productive)
Internal to firm			
Own R&D	33	13	5
Other internal sources (e.g., engineering or maintenance)	41	45	40
Other firm in group	28	9	1
The market			
Clients or consumers	50	38	34
Competitors	19	21	22
Technology market inputs			
Suppliers (equipment)	30	40	34
Acquisition of licenses, patents, and know-how	8	4	2
Consulting firms	10	8	3
Specialized technology support infrastructure			
Universities and research labs	8	7	4
Professional training & technical assistance centers	8	7	5
Metrology and testing & accreditation centers	12	8	5
Sources of technological information			
Conferences and publications	17	14	15
Fairs and expositions	33	37	33
Information networks	24	17	13

Source: Based on data from Koeller and Baesa, "Inovação tecnológica na indústria Brasileira," in IPEA (2005).

Table 5.6. Obstacles to Innovation for Brazilian Firms, 1998–2003

percent of firms with at least 10 employees

Main obstacles to innovation	1998–2000	2001–03
Costs	82.8	79.7
Economic risk	76.4	74.5
Scarcity of financing	62.1	56.6
Shortage of skilled labor	45.6	47.5
Lack of information	36.6	35.8
Difficulty adopting standards	25.1	32.9

Source: Cruz and De Mello 2006, based on PINTEC.

Relationship among Innovation, Productivity, and Growth

For decades, analyzing and quantifying the effects of innovative activities on productivity has been a challenging and controversial task in empirical economics (Janz et al. 2003). In the 1990s research on this topic was enriched by new theoretical underpinnings from endogenous growth theory showing that economic output should be positively related to the flow of innovations.[4] In the case of Brazil, firm-level studies recently have been developed, based on data from the PINTEC survey. The ICS data, collected by the World Bank, permits further exploration of these topics. Findings from these analyses are summarized below.

Findings from the PINTEC Database

To explore the relationship between innovation and exports and the performance of manufacturing firms in Brazil, we used Arbache (2005).

The initial and final periods of this cross-section analysis are 1997 and 2001, respectively. The econometric models divided firms into the three categories already noted in the Institute of Applied Economic Research (IPEA) data. One productivity measure used in the analysis was the log of potential value added per worker (log PVA per worker), measured as the log of value added (total net sales less operational costs minus total wages divided by the number of workers). Results of this exercise are presented more fully in appendix B.

In brief, the appendix B results show that innovation through new product development boosts firm productivity: a company that introduces new technological products to the market has productivity 23 percent higher than a company that does not innovate. Regarding R&D intensity (that is, R&D expenditures as a share of total sales), increasing returns to scale were found, which probably were associated with the initial development stage of firms' R&D investments in Brazil. In addition, a 1 percent increase in R&D intensity would be associated with an increase of 0.2 percent in the firm's productivity—and almost 0.5 percent for firms specializing in standard products.

Exporting also was found to be associated with higher productivity: exporters have productivity 161 percent higher than nonexporters. In addition, a 1 percent rise in exports as a share of total sales would be associated with a 13 percent jump in productivity. For firms specializing in standard products, this elasticity was only 7 percent, while the productivity of firms that innovate and differentiate their products does not change when exports increase as a share of total sales.[5]

The education of the labor force also was related positively with productivity. A 1 percent increase in the average education of the labor force would be associated with an increase of 0.63 percent in productivity. This elasticity was 1.29 percent for firms specializing in standard products, which implies that investments in human capital present increasing returns to scale. Finally, multinational firms would have higher productivity than purely domestic enterprises.

At the firm level, evidence is strong on the positive relationship between R&D, innovation, and productivity. However, causality cannot necessarily be inferred because these are cross-section data. In order to assess the causality between innovation and a firm's performance (after auto-selection treatment), a counterfactual exercise was developed. For example, results for cluster one (eight clusters were created) showed that the growth rate of firms that innovated in both 1997 and 2001 was 6.28 percent, while it was only 0.46 percent for firms that innovated in 1997 but did not in 1998–2000, resulting in a difference of 5.82 percent. This suggests that the employment rate of firms that stopped innovating after 1997 grew more slowly than that of their counterparts who kept on innovating. Considering all eight clusters (appendix B, table B.2), results for Brazil show that innovation *causes* increased firm size in terms of higher employment and improved productivity.

Findings from the Investment Climate Survey (ICS) Database

With more than 1,600 firms included in the sample for Brazil, the ICS database also allows for investigating the relationship between productivity and investment climate (IC) variables related to technology and innovation. Two exercises were undertaken to explore this relationship.

First, Escribano et al. (forthcoming) did a cross-country comparison to determine how a set of IC variables affect manufacturing sector productivity in seven countries through a change in TFP and two other competitiveness indicators: the probability of a firm exporting and its probability of receiving FDI resources. The dataset was composed of the ICS data for Brazil, Ecuador, El Salvador, Guatemala, Honduras, and Nicaragua, as well as one Asian benchmark, Indonesia. Data were pooled from 4,679 firms, representing nine manufacturing sectors.

Using an econometric model based on Escribano and Guasch (2004), a two-step estimation was developed: (a) estimating the parameters of a panel data regression model by pooling observations from several countries to get a large sample size for consistent and asymptotically efficient estimators, and (b) evaluating the impact of each IC variable in the sample means by using two-stage least squares to compute the impact on "average productivities."[6] What followed was a country-by-country evaluation of the impacts of IC variables on competitiveness indicators.[7] Hence, comparisons across countries are not as robust in their specification as, for example, in the estimation described above.

Highlighted findings for the technological variables include the following:

- *Effects on productivity for the pooled multicountry data*—International Standards Organization (ISO) certification and worker training have a statistically positive impact on the probability that a firm exports and has shares of foreign ownership. These results were robust for both TFP measures (that is, restricted and unrestricted by industry cases).[8] In addition, computer use (measured as the share of workers using computers) and Web use by the firm (dummy) have a statistically positive impact on both TFP measures.

- *Effects on productivity at the country level*—among the six Latin American countries, greater computer use by workers and company Internet access would have the largest impacts on TFP.

- *Effects of improving certain IC variables by aligning them with top performers*—for Brazil, three IC variables would have the greatest positive impact on productivity. They are (a) the average time to clear customs (13.6 percent), (b) lost sales from transport interruptions (4.9 percent), and (c) the share of workers using computers (1.3 percent).

Another exercise for the IC assessment was carried out with Brazilian data only (table 5.7). The econometric analysis of the determinants of TFP shows that IC variables related to innovation and technology adoption are statistically significant.[9] To highlight the differences of IC effects on enterprises of different sizes, the analysis of the whole sample was repeated for two subsamples—micro and small enterprises (MSEs) and medium and large enterprises (MLEs). Results show that innovation, skills, and quality standards are important determinants of TFP. The use of computers by workers, the acquisition of an International Standards Organization (ISO) certificate, the manager's education level, and the provision of external training to workers are significant factors behind higher TFP. For example, the average TFP differential for plants in which the general manager does (versus does not) have some college education is around 20 percent. The provision of external training is especially important for smaller firms, resulting in an 11.5 percent increase in TFP if everything else is held constant. For larger firms, the acquisition of an ISO certificate is associated with a 17.4 percent increase in TFP. Finally, if the share of workers using computers increases by 1 percent, the increase in TFP would be 0.5 percent for MSEs and 1 percent among MLEs.[10]

Firm-Level Analysis of the Relationship among R&D, Innovation, and Productivity[11]

A more complete exercise using the ICS data was carried out by Correa et al. (forthcoming), who simultaneously modeled the determinants of R&D, innovation, and productivity to understand the channels linking investment in knowledge and innovation to productivity growth at the firm level. This kind of analysis addresses several questions: Does the level of engineers affect R&D, innovation, and productivity? Does firm size matter? How does innovation compare with technology adoption in affecting productivity?[12]

In this exercise, two analytic models were combined: (a) one developed by Crepon, Duguet, and Mairesse (1998) for R&D, innovation, and productivity that explicitly models, in a simultaneous equation framework, the path by which investment in research generates knowledge and the forms by which such knowledge is transformed into outputs; and (b) the methodology developed by Escribano and Guasch (2004) for estimating productivity by incorporating IC variables.

Table 5.7. Average Coefficients (Semi-Elasticities) for Selected Investment Climate Variables Estimated from TFP Regressions for Brazil

Dependent variable in the production function: log (value added)	Sample		
	All	MLE	MSE
Red tape, corruption, and crime			
Loss due to theft (% sales)[a]	−3.3	−5.1	−3.0
Delays of imports in customs (average days)[a]	−0.7	−0.3	−1.3
Senior management's time spent on regulation (%)[a]	−0.3	−0.5	0.0
Infrastructure			
Power interruptions (index)[b]	−12.9	−14.9	−11.8
Communications interruptions (index)[b]	0.0	−17.0	0.0
Transportation interruptions (index)[b]	−12.8	−31.4	0.0
Innovation, quality, and skills			
Staff using computers (%)[a]	0.9	0.5	1.0
ISO certificate (vs. no ISO certificate)[c]	9.2	0.0	17.4
General manager with at least some college education (vs. not)[c]	21.2	20.8	21.8
External training offered (vs. not)[c]	11.5	3.4	11.5
Labor regulation			
Informal workers among full-time employees (%)[a]	−0.3	0.0	−0.3
Finance and corporate governance			
Needed bank loan, but did not apply (vs. applied and failed)[c]	15.9	0.0	15.7
Apply external audit (vs. not)[c]	2.2	3.3	11.7
Other			
Inputs imported (%)[a]	0.2	0.0	0.3

Source: World Bank 2005a.

Note: All the coefficients have been premultiplied by a factor of 100 to reflect the impact on TFP in percentage terms.

a. The coefficient for this variable can be interpreted as the change in TFP (%) corresponding to a one percentage point (or one day) increase in the value of the variable.

b. The coefficient for this index can be interpreted as the change in TFP (%) corresponding to a one point increase in the value of the index (the index ranges from 0 [best] to 4 [worst]).

c. The coefficient for this (dummy) variable can be interpreted as the average TFP gap (%) between the plants in the category versus the plants that belong to the group identified in the brackets.

The econometric model comprises three sets of equations that are estimated together and reported in appendix B (table B.1 and table B.2).

R&D Determinants. The firm's decision on whether or not to engage in R&D was determined primarily by firm size, credit access,[13] and the availability of qualified personnel. Once the firm decided to invest in R&D activities, the determinants of R&D expenditures per worker were firm size and market share. After controlling for market share, the well-documented positive relationship between size and R&D expenditures per worker does not hold; in fact, the intensity of R&D activities decreases with firm size. The coefficient of size in the model is the size elasticity of R&D expenditures per worker (–0.3), which means that if the size of the company doubles, R&D expenditures per worker decrease by 30 percent.[14]

These findings also are observed in other countries and confirm the Schumpeterian hypothesis that R&D is undertaken mostly by large monopolistic firms (Schumpeter 1942). This implies that market dominance is necessary to undertake the risks and uncertainties associated with R&D. Furthermore, studies have found that market power is more important than the absolute firm size for reaping the benefits of innovative activities (Cohen et al. 1987); hence, only enterprises that are large enough to secure at least temporary market power will innovate.

Innovation Determinants. Slightly different results were found in the two versions of the innovation equation. After controlling for industry and region, the availability of qualified personnel was found to be the only significant determinant of a firm becoming an innovator. On the other hand, R&D expenditures per worker and firm size are positively related to innovation intensity. For instance, a 10 percent increase in R&D expenditures per worker would be associated with an approximate increase of 4 percent in the enterprise's portfolio of new products. Increases in firm size induce effects in opposite directions—a direct positive effect on innovation intensity and an indirect negative effect on reducing R&D expenditures per worker. The overall net effect is negative: a 10 percent increase in employment with no adjustments in R&D expenditures per worker would be associated with a 1.3 percent decrease in innovation intensity.

In sum, these results point to two distinct channels for becoming established in the market as an innovative firm. One channel is through more sophisticated inventive activities—that is, R&D in the conventional sense. This kind of activity may or may not translate into salable products, but once it does, a firm is well positioned to widen its portfolio of products in the marketplace. The second channel is through high-skilled workers who are able to transform existing technologies into new products. This channel does not necessarily require technological sophistication.

Productivity Determinants. The results of the joint estimation show that innovation (in both specifications) positively affects a firm's productivity (measured as value added per worker). In the case of innovation intensity, a 10 percent increase in a firm's portfolio of new products would be associated with a 5 percent increase in value added per worker. Coefficients associated with computer use and quality certification also are related positively and significantly with productivity. These two variables can be interpreted as proxies for technology adoption by the firm and managerial capabilities, respectively. For instance, a 10 percent increase in the share of workers using computers (in absolute terms) is associated with a 12 percent increase in productivity (in relative terms). In addition, firms holding an ISO certificate are 30 percent more productive than those without it.

Large productivity gains were also seen in publicly listed firms: their value added per worker would be 48 percent higher than the productivity of firms presenting a different legal status. Capital stock and capital use are also relevant determinants of productivity. For example, a 10 percent increase in the firm's capital stock would be related with a 3 percent increase in value added per employee.

Although the signs for most of the IC coefficients generally turn out to be positive as expected in the joint estimation, the relationships do not all turn out to be statistically significant. A positive relationship between size and R&D, innovation, and productivity is typical. By including size in the three equations, it was possible to test whether size would have a significant direct effect in each equation—in other words, whether its effect would disappear once R&D is controlled for in the innovation equation or in the productivity equation. In both specifications, the findings show that firm size is an important determinant of R&D and innovation for Brazilian manufacturing firms. However, results for productivity differ depending on whether innovation is measured as a dummy variable or as innovation intensity. In the first case, size still plays a role in explaining value added per worker; but when innovation output is measured by innovation intensity, the size effect disappears.

Analysis of Inputs and Outputs Related to Innovation by Key Variables

This section presents evidence from the Brazilian ICS on firms' inputs and outputs related to innovation. The analysis is broken down by firm size, sector, location, ownership, and exporting status. Key findings are briefly reported. Fuller analysis is provided in World Bank (2005a).

First, innovation inputs and outputs in Brazil are positively related with firm size (figures 5.1 and 5.2). Significant variation, however, occurs across industrial sector and region. For instance, among large firms (more than 500 workers), 74 percent report R&D spending. This compares with 39 percent among microenterprises, 48 percent among small firms, and 60 percent among medium firms. Similar differences were found for other innovation inputs (technology licensing, joint ventures, and worker training) as well as for innovation outputs (new products and improved product lines).

Second, when industrial sectors are considered (figures 5.3 and 5.4), the largest percentage of firms investing in innovation inputs is found to be in electronics, machinery, and auto parts (the sectors, not surprisingly, with the highest average of foreign ownership). By contrast, leather and footwear, and apparel are the two sectors with the lowest percentage of firms investing in innovation inputs (particularly in ISO certification, joint ventures, and technology licenses). However while auto parts had the highest rate of new products, it was followed closely by footwear, furniture, and food—all of which had higher rates of product innovation than did electronics or machinery. This again highlights the point that R&D is not necessarily the key to product innovation.

Figure 5.1. Innovation Inputs (R&D, Licensing, Joint Ventures, and ISO) by Firm Size

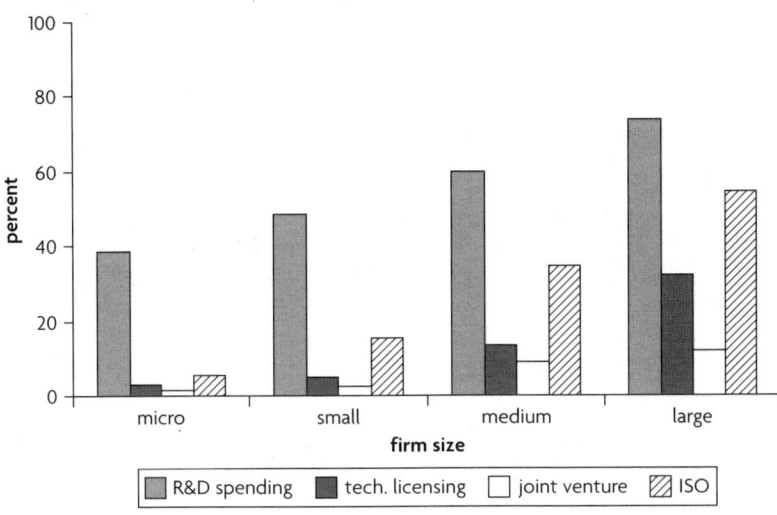

Source: Brazil Investment Climate Assessment 2005.

Figure 5.2. Innovation Outputs (Training, Improved Line, and New Product) by Firm Size

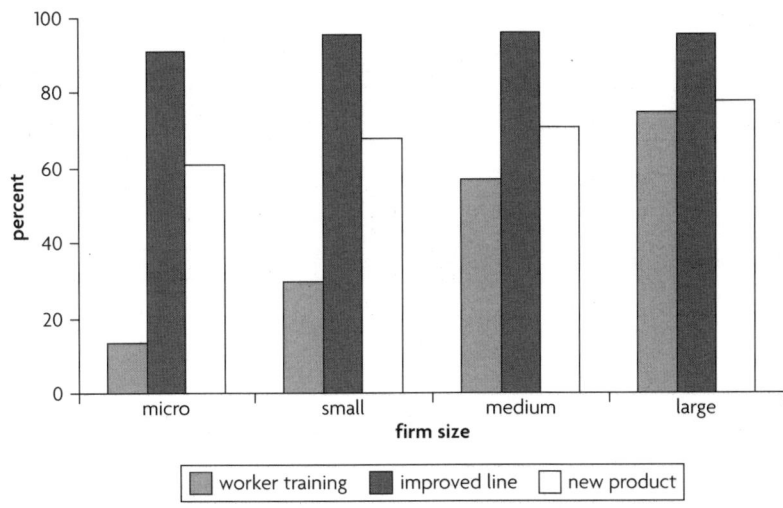

Source: Brazil Investment Climate Assessment 2005.

Third, when disaggregating by region (figures 5.5 and 5.6), the largest percentage of firms investing in innovation inputs and outputs is found in the
south of Brazil, with the exception of joint venture agreements (the largest
percentage of firms holding these agreements, 5 percent, is in the southeast).
The lowest percentage of firms investing in innovation inputs and outputs is
in the northeast (with the exception of ISO certificates, for which the center
and west show the lowest shares).

Figure 5.3. Innovation Inputs (R&D, Licensing, Joint Ventures, and ISO) by Sector

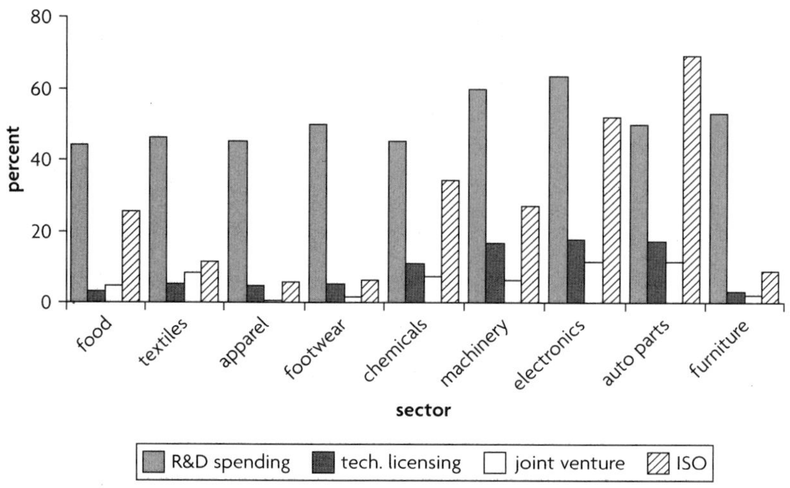

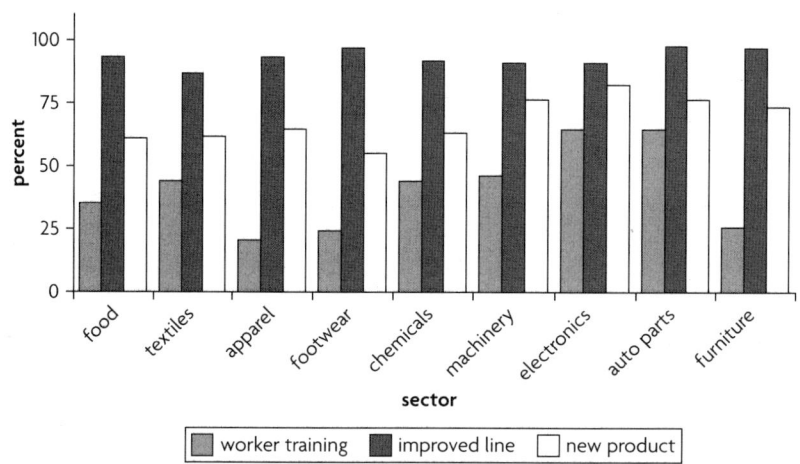

Source: Brazil Investment Climate Assessment 2005.

Figure 5.4. Innovation Outputs (Training, Improved Line, and New Product) by Sector

Source: Brazil Investment Climate Assessment 2005.

Appendix C uses Probit estimation to examine these relationships in greater depth. Appendix tables report marginal effects; so it is possible to assess the magnitude of the partial effects associated with changes in the explanatory variable for each dependent variable. In brief, we find that the effects of firm size persist when controlled simultaneously in a regression framework. This holds true even when controlling for sector and location. It is also interesting to note that exporting firms and those supplying foreign-owned companies also generally reveal a greater incidence of innovation activities. For instance,

Figure 5.5. Innovation Inputs (R&D, Licensing, Joint Ventures, and ISO) by Region

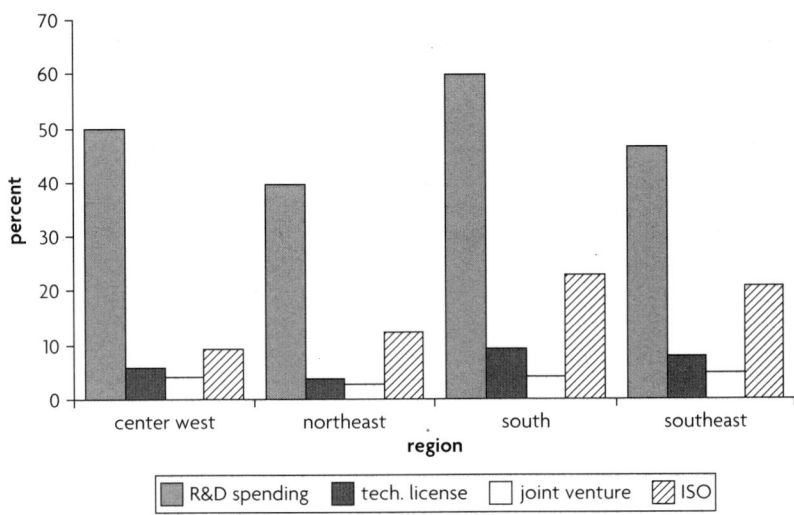

Source: Brazil Investment Climate Assessment 2005.

Figure 5.6. Innovation Outputs (Training, Improved Line, and New Product) by Region

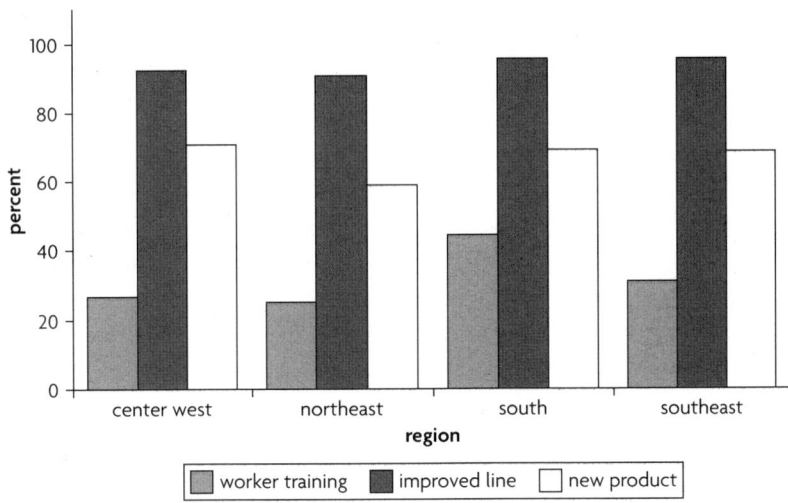

Source: Brazil Investment Climate Assessment 2005.

a 1 percent increase in sales to foreign firms as a share of total sales would increase the firm's probability of investing in R&D by 0.1 percent, holding an ISO certificate by 0.1 percent, providing worker training by 1.2 percent, and developing new products by 0.05 percent.

Similarly, innovation inputs and outputs also are more likely to occur among firms that employ more educated workers (table 5.8). Firms with a higher share

Table 5.8. Marginal Effects of Education on Innovation Inputs and Outputs in Brazil

Independent variables	R&D	ISO	Worker training	Joint venture	Tech. licenses	New product	Improved line
Employees with	0.001**	0.001**	0.002***	−0.001*	0.001*	0.002**	0.001***
high school (%)	[1.96]	[2.17]	[4.11]	[1.70]	[1.64]	[2.52]	[3.18]
Employees with	0.008***	0.004***	0.006***	−0.001	0.001	0.004***	0.001**
some college (%)	[5.13]	[4.14]	[4.14]	[0.18]	[0.73]	[2.68]	[1.99]
Observations	1,631	1,554	1,630	1,631	1,631	1,631	1,631
LR χ^2 (d.f. = 21)	161.83	566.68	462.60	117.13	207.69	91.06	63.80
Pseudo R^2	0.072	0.375	0.221	0.210	0.240	0.044	0.092

Source: Brazil Investment Climate Assessment 2005.
Note: Z-value is in brackets. For brevity, variables for sector, region, size, export status, foreign ownership, share of sales to exporters, and share of sales to foreign-owned firms were not included. Wood and furniture is the omitted category for sector. Southeast is the omitted category for region. Micro is the omitted variable for size.
*Significant at the 10 percent level.
**Significant at the 5 percent level.
***Significant at the 1 percent level.

of employees who have completed secondary and some college education are more likely to invest in R&D, hold an ISO certificate, provide worker training, develop new products, and improve new lines of production. For example, if a firm increases its share of high-school-graduate employees by 10 percent, its probability of investing in R&D increases by 0.01 percent, everything else held constant; the probability increases by 0.08 percent if the share of employees with some college education is increased by 10 percent. Moreover, the larger the share of employees who are high school graduates, the more likely the firm is to acquire technology licenses, provide worker training, develop new products, and improve its line of production (the opposite impact was found on joint venture agreements).

In summary, we can draw four conclusions on the general relationships between the characteristics of Brazilian manufacturing firms and their likelihood of being innovators. These have important implications for the broader questions posed in chapters 1 and 2, and for forward-looking strategies discussed in the final chapter.

First, size matters for innovation. Using discrete size categories, results show that small, medium, and large firms have higher (and increasing) probabilities of investing in innovation inputs and developing outputs than do microenterprises. These findings confirm the consensus that R&D activities and innovation increase as the size of the firm increases.[15] Capital market imperfections as a source of competitive advantage for large firms are confirmed as a main argument for sustaining the relationship between firm size and innovation.

Second, exporting is also a determinant of innovation for Brazilian manufacturing firms. Exporters have higher likelihoods of investing in innovation inputs/outputs than nonexporters. It can be argued that exporters can often access diverse knowledge inputs unavailable in the domestic market, that this knowledge can spill back to the local firm, and that such learning can foster innovation.

Third, foreign ownership matters for innovation. Brazilian manufacturers with some degree of foreign ownership are more likely to innovate than purely domestically owned firms. Foreign ownership has a positive effect on innovation because of the resources (finance, technology, knowledge, and managerial expertise, for example) that foreign parties are able to tap for their Brazilian holdings, which cannot necessarily be reproduced by smaller Brazilian-owned firms. The foreign-ownership effect captures the manner and the extent to which an overseas shareholder is able to add value to the domestic firm and reduce barriers to the local development of innovative activities.

Fourth, human capital is also found to be a significant correlate of innovation among Brazilian manufacturing firms. This confirms the hypothesis that human capital is complementary to innovation and technological change.[16]

Human Capital, Innovation, and Productivity

As noted in the marginal analysis reported in table 5.8, human capital is a significant correlate of innovation among Brazilian manufacturing firms. This positive relationship between human capital and innovation inputs as well as outputs also has been found in many parts of the analysis noted above. Because this relationship is a primary focus of this report, we provide additional evidence. The first piece is in table 5.9, which shows the average years of schooling for the different firm groups in the PINTEC data. In firms characterized by competitive strategies based on innovation and product differentiation, average schooling is almost two years higher than for firms characterized as specializing in standardized products. The average for years of schooling by workers in the second category, in turn, is almost one more year than the average for workers in firms that do not differentiate their products and have lower productivity. The average job tenure also is correlated strongly with years of schooling, suggesting that the more innovative firms also probably invest more in worker training and retrain longer.

Table 5.9. Average Wages, Schooling, and Worker Tenure in Brazilian Firms by Competitiveness Group Type in 2000

	Average wages (R$/month)	Average years of schooling	Average month on the job	Wage premium (%)
Group A: Innovative and product-differentiating firms	1,254	9.13	54.09	23
Group B: Firms specializing in standard products	749	7.64	43.90	11
Group C: Firms that do not differentiate products and are less productive	431	6.89	35.41	0

Source: Bahia and Arbache, "Diferenciação salarial segundo critérios de desempenho das firmas industriais Brasileiras," in IPEA (2005).

A second and more compelling piece of evidence comes from recent econometric work on worker characteristics and technology absorption in Brazilian industrial firms by F. de Negri (2006). That work explicitly used the PINTEC database to analyze the probability of innovation by Brazilian firms and the extent to which external sources of information were used to innovate. Findings were robust to various specifications, including a multinomial Probit model. The study found that both the technological effort of firms and the level of worker schooling were statistically significant determinants of innovativeness by Brazilian firms. With regard to the relationship between workers' characteristics and the absorptive capability of the firm, table 5.10 shows that by far the most significant determinant of a firm's absorptive capability was the percentage of its workers with higher education, followed by whether it had a formal R&D department (continuous R&D effort). The number of workers in the firm and the broadness of training also were related positively

Table 5.10. Probit Model of Probability Factors for Absorbing Technology by Brazilian Firms

Explicatory variables	Est. coefficient	Standard deviation	Marginal probability
Intercept	0.616	0.293**	0.234
Occupied employees (natural log)	0.058	0.009***	0.220
Dummy for firm w/ continuous R&D	0.189	0.028***	0.072
R&D expenditures as proportion of sales	0.002	0.000***	0.001
Dummy for firm w/ staff training (1999)	−0.150	0.040***	−0.057
Average employment time in 1997	−0.005	0.001***	−0.002
Average employment time in firms w/ staff training	0.004	0.001***	0.002
Staff w/ higher education in 1997 (%)	0.671	0.132***	0.255
Herfindahl-Hirschman Index (1997)	−0.173	0.040***	−0.065
Average work experience of firms' employees	0.005	0.002**	0.002
No. of cases in sample	5,042	$L_0 = -16{,}435$	
No. of firms w/ absorptive capacity (population)	7,755	$L_1 = -14{,}108$	
No. of firms w/o absorptive capacity (population)	15.006	Pseudo $R^2 = 0.14$	

Sources: De Negri (2006), with Probit model estimated from PINTEC (2000) database and RAIS (1997) database.
**Significant at the 5 percent level.
***Significant at the 1 percent level.

to innovation (the opposite of the Herfindahl-Hirschman Index, a commonly accepted measure of market concentration, which had a negative sign). Somewhat surprising was that the length of worker tenure and a dummy for worker training had negative signs. However, the length of worker training in firms that did provide training had a positive impact. The author suggests that on-the-job experience without training may not contribute to absorption and that training may have no impact if there is high labor turnover. The study also found that greater education was necessary to utilize academic information sources rather than industry information sources. This implies that increasing the workforce educational level, particularly the percentage of workers with tertiary education, is an important factor in boosting firms' absorptive capability. This would be relevant for absorbing technology from universities and research institutes as well as from multinationals, other firms, and suppliers.

Conclusions and Policy Implications

This chapter explored data from PINTEC and the ICS database. Several findings emerged:

- Innovation is less intense in Brazil than in OECD countries. In addition, innovation in Brazil is more prevalent in processes than in new products.

- Innovation was found to be important for productivity and for growth. Therefore, much more needs to be done to stimulate innovation in the Brazilian economy.

- Innovation is not confined to R&D and often occurs without any. R&D, however, can play an easily overlooked role in acquiring and using technology, whether domestically or from abroad. Understanding that the most important source of innovation is equipment and machinery is particularly important in Brazil, given the country's low private investment rate compared with its competitors. That handicap is compounded by the finding that Brazil imports relatively few capital goods, even compared with competitors with equally or more developed capital goods sectors. Brazil not only must increase its investment rate; it needs to further liberalize tariff and nontariff restrictions on capital goods imports.

- Brazilian firms also were found to be less likely than their competitors to take advantage of establishing joint ventures and licensing foreign technology. Among the various modes of technology adoption, Brazilian firms rely surprisingly little on international technology transfer. This is at odds with the increased importance of the international trade of goods and services in facilitating knowledge absorption. Indeed, foreign sources of technology account for 90 percent of technology transfer in most countries, while the bulk of R&D—an indication of new knowledge creation—is concentrated in a few countries (Keller 2004). Imports of intermediate inputs, machinery, and equipment are critical channels of technology transfer. One factor behind Brazil's relative lack of success in acquiring technology, therefore, may be its relatively poor integration into the global trading system. In fact, Brazil's trade volume is low even when compared with large countries such as China and India. However, even when the volume of trade is controlled for, Brazilian capital goods imports are below the international average. But low imports of capital goods also may be related to relatively higher sector-specific tariff barriers and to the availability of financing for the acquisition of local equipment. Or there may be structural barriers, ranging from the lack of appropriate logistical services to the poor education level of the labor force.

- In Brazil, firm size, exporting status, foreign ownership, and human capital matter for innovation, even when controlling for region and industrial sector. Findings were robust and stable. Results show that small, medium, and large firms have higher (and increasing) probabilities of investing in innovation inputs and developing outputs than do microenterprises. Exporters and firms with some degree of foreign ownership have higher likelihoods of investing in innovation inputs/outputs than nonexporters and domestically owned enterprises. Finally, human capital also is found to be a significant correlate of innovation inputs and outputs.

- The econometric evidence therefore reinforces the point that technology adoption and R&D do matter, but the contribution of R&D to productivity seems smaller than the contributions from technology adoption and other aspects of the investment climate. Both innovation measured as a dummy and innovation intensity depend on R&D expenditures and the supply of skilled labor. However, these are relatively expensive activities and may be limited by other investment climate variables.

- This does not mean that R&D is not relevant for long-term growth or that Brazil should not pursue an R&D policy. However, data suggest that, given Brazil's development level, some emphasis on technology adoption (international transfer and local diffusion) may be more cost-effective than R&D at the firm level. The impact of technology adoption on labor productivity is particularly evident from the results obtained for the capital stock and computer-use variables in the productivity equations.

- Technology adoption varies with firm size and industry, which contrasts with the Brazilian S&T policy emphasis on R&D support to capital-intensive industries and, most likely, large firms. A broader set of measures to support technology adoption for SMEs in labor-intensive industries is still missing. This also may help accelerate the process of technology diffusion. Recent steps were taken in this direction (Law No. 11,196/2005 and the Statute of SMEs), yet still more needs to be done. Based on these findings at the firm level, more also must be done to promote technology diffusion, including implementation of technological information systems, technology extension, and demonstration projects and the upgrading of worker skills.

- Other investment climate issues—including infrastructure, competition, and business regulation—also affect technology adoption. Econometric analysis suggests that investment climate variables, such as infrastructure, limit the impact of technology adoption and innovation on productivity. Rather than increasing public R&D expenditures, it would be more fruitful to understand why the private sector has been less active in this area and to remove the bottlenecks to greater private R&D investments or efforts to innovate. Addressing the broader investment climate constraints like access to capital, labor market rigidities, and property rights could be a more effective approach and is likely to be the most successful way of enhancing the incentives for firms to innovate, create jobs, and grow.

- Finally, ample empirical evidence underlined the importance of education and skills in absorbing existing technology, whether obtained locally or from abroad, and in creating knowledge. Brazil is weak in this kind of human capital compared with its competitors. The next chapter will analyze why Brazil is falling short and what it can do to catch up.

CHAPTER 6

Human Capital for Innovation and Growth

A firm that utilizes advanced technologies tends to employ better-qualified workers who understand and can operate the technologies. At the same time, qualified workers are able to improve the technological performance and competitiveness of the firm, thus contributing to its creative potential.

—De Negri et al. 2006, p. 374 (author's translation)

Previous chapters showed how Brazil's manufacturing output and productivity might be increased if more firms adopted innovation-enhancing technologies already present in the country. This chapter explores how weakness in the education system has been one of the key shortcomings preventing this from occurring.

Brazil has made important strides in education in recent years—particularly in relation to school access and equity. However, other countries have, too. And in comparison with these countries, Brazil appears to be underperforming at every level, from preschool to postdoctoral research training. From the point of view of a highly competitive global economy, Brazil's education systems are failing to create an innovation-ready workforce.

This chapter examines the nature of human capital—the missing link between innovation and productivity—compares the formation of human capital in Brazil with that of its competitors, and assesses the changing demand for skills in Brazil's job market. It then provides summary tables that profile Brazil's education system at every level, including advanced skills training outside the formal education system. Two detailed appendixes—the Primary Education System (appendix D) and the Tertiary Education System and Advanced Out-of-School Training (appendix E)—are included at the end of this report. These appendixes explore specific educational issues in greater detail and provide a fuller portrait of the various educational components summarized in this chapter.

Jamil Salmi and Domenec Devesa were key contributors to this chapter.

Human Capital: The Missing Link between Innovation and Productivity

With the transition from the industrial economy of the 20th century to the knowledge economy of the 21st century, the global marketplace increasingly has rewarded flexible, efficient economies that are able to rapidly adapt to new circumstances—in a word, those that can "innovate." Countries that have been successful most recently are those that have mechanisms in place for expanding trade, producing knowledge, and putting technology to efficient use. Increasingly, these countries participate in the global chain through economic conversion toward higher value-added activities.

As discussed in previous chapters of this report, despite its successes, Brazil is not yet fully prepared to compete in this new global environment. Its economy is still based heavily on primary commodities and exploitation of natural resources. Its trade policies remain protectionist. Too often its labor laws hold back formal employment. Bureaucratic red tape, high taxes, and high interest rates discourage firms from investing. Taken together, these characteristics create an investment climate that hinders rather than enables manufacturing firms' ability to "plug in" and find new niches for higher growth.

As argued in our conceptual framework (chapter 2), higher productivity can be achieved along three pathways—through increases in physical capital, human capital, or TFP (that is, gains through greater efficiency in how physical and human capital interact—basically what we are calling "innovation"). Chapter 3 argues that there are three broad types of innovation—advances that arise from *creating* new knowledge and technology, those that come from *acquiring and adapting* new knowledge and technology from abroad, and finally those that come from *absorbing and using* existing in-country knowledge to improve processes and products. In general, improvements in the effectiveness of public R&D stimulate creation of new knowledge, while expanding private sector investment stimulates acquisition and absorption of new knowledge. Yet investment in either sort of innovation does not automatically lead to greater productivity. Something else is needed—innovation-ready human resources—to "add the value" made possible through either kind of investment. If this link is missing, nothing else happens.

Human Capital and the Three Categories of Innovative Activity

Important evidence shows how human capital affects innovation activity—and in particular, the degree to which public education shapes the supply and capacity of workers to innovate. It is now widely agreed that technological change is linked to the supply of highly educated workers (World Bank Institute 2006). Moreover, while other paths to forming human capital exist, such as on-the-job-training, the evidence presented in chapter 5 shows that firm-level innovation is highly dependent on the education received by employees in a primarily public education system.

So what do we know about the contribution of human capital to innovation in Brazil? As discussed above, our broad definition of innovation includes creating new knowledge as well as acquiring and adapting knowledge from abroad and absorbing and using in-country existing knowledge. Each must be understood as a different, but important, process of innovation.

Creating New Knowledge and Technologies. The first kind of innovation results from the creation of new knowledge and technologies, either for the domestic or sometimes the international market. Creation of these new technologies and knowledge (usually packaged as "new products") requires a whole host of antecedent activities, ranging from R&D and market research to engineering the necessary production facilities (De Negri et al. 2006). Because of the complex process involved, this innovation type predictably requires workers with higher levels of schooling and extensive job training.

Acquiring and Adapting Foreign Technologies. Acquiring knowledge and technologies from abroad most commonly involves importing new technologies (typically machinery), which may require significant upgrades of worker capacities. Thus, a critical factor in such innovation is workers' ability to learn to operate new machines through reading and understanding product manuals and learning to provide routine maintenance through basic training. Equally important is the ability of process planners and supervisors to resolve problems in adapting equipment to its new setting and revising processes and procedures to take full advantage of the potential it offers. Obviously, for this type of innovation, both basic skills for operators and advanced skills for supervisors and planners are critical.

Using, Adapting, and Disseminating Existing Technology. The third kind of innovation involves the absorption and use of knowledge and technologies already in the country, which implies diffusion of a technology or process already being used by harnessing it elsewhere. These less obvious forms of innovation also require the upgrading of existing human capital stocks through on-the-job learning. Basic skills in reading, communication, and mathematics are critical because they are the springboard for further learning that ultimately culminates in process and product improvements.

Levels of Worker Capacity Relative to Categories of Innovative Activity

Whichever type of innovation is involved, human capital and worker skills are clearly a critical input for fueling innovation, productivity, and competitiveness. An argument for the payoffs to be earned from raising worker capacities can be made by observing the spread of educational levels among firms divided into different innovation categories. Table 6.1, for instance, shows Brazilian industrial firms with innovative product lines have workforces that average

Table 6.1. Salaries, Schooling, and Tenure in Brazilian Industrial Firms

Competitive strategy classification	Wage average (R$/month)	Average schooling of workforce (years)	Average tenure (months)
1. Firms competing through product innovation and differentiation	1,255	9.1	54.1
2. Firms competing through cost-cutting strategies	749	7.6	44.0
3. Firms with difficulty competing because of low productivity	431	6.9	35.4

Source: Arbix (forthcoming); based on data from IBGE's 2000 Technological Innovation Survey (PINTEC, *Pesquisa de Inovação Tecnológica*).

9.1 years of education, compared with only 6.9 for workforces in firms that are described as non-innovative (Arbix, forthcoming). Moreover, employees in the first category earn more and have greater job stability.

Confirming the above-mentioned relationship, econometric analysis based on the firm-level 2005 Investment Climate Survey shows a negative correlation between innovation and the proportion of workers with only primary education (World Bank 2007b). Another dichotomy exists between exporting and non-exporting firms. The differences in workforce educational levels are noticeable, as exporting firm employees have 9.9 years of schooling versus 8.5 for non-exporter employees (De Negri et al. 2006). Interestingly, robust econometric evidence shows a strong negative correlation between innovation and in-house firm training in Brazil (World Bank 2007b). This probably stems from the training's focus on compensating for the basic skill deficiencies of the workforce. In other words, training is used to level the playing field rather than to expand the productivity and efficiency horizon of workers. Obviously, a stronger educational system providing solid fundamental skills would allow firms to refocus their resources and training on upgrading specific technical skills that could drive greater innovation in the workplace.

In summary, there is clearly a link between human capital and innovation, but once again, this relationship needs to be broken down into more specific components to translate findings into public policies. Having more engineers with better qualifications argues for expanding access to higher education and emphasizing certain kinds of careers. But links between innovation and average worker education on the shop floor emphasize the importance of basic education.

Keeping these factors in mind, the discussion now turns to Brazil's performance in developing skills among its populace. To contextualize the issues, we begin by describing the changes in requisite employment skills that have occurred, and then we compare and contrast the Brazilian experience of human capital development with that of key competitors in the global market. Finally, the Brazilian education system is given close scrutiny as we examine its progress and shortcomings, and appraise the challenge it faces in providing the basic and advanced skills needed by future generations of knowledge workers.

Changing Demand for Skills in Brazil's Job Market

In an earlier section we reviewed evidence suggesting that Brazil, to enhance its innovation practices, needs to invest more in education and training and create a more dynamic workforce—and workplace. But acting on these needs is another matter. For example, what kinds of skills do workers really need? We review some evidence about this below. Moreover, we have already made the functional case that more innovative firms in Brazil have higher levels of worker human capital based on comparisons of education levels among different firms. Indeed, does the labor market reward education level? Some of these issues are addressed by De Ferranti and Perry (2003), who examine education and technology in the Latin American and Caribbean region. Their evidence is consistent with a global pattern in which the highest wage premiums are associated with higher education credentials and, presumably, are rewards for individuals who are able to help firms harness the kinds of rapid technological change that are necessary to compete. Table 6.2 presents the average salaries paid in Brazil (in 2004) according to the level of education of the employee.

Clearly, these salary differentials are not only due to skills but are also determined by other factors such as the relative supply of workers, the growth rates of these supplies, the industry, and other enabling conditions. But even with the uncertainties generated by rapidly changing enabling conditions, the job market clearly is rewarding those with higher levels of education. The likely reason is that those with higher levels of education are able to learn and adapt quickly to changing situations. And as we next discuss, the knowledge

Table 6.2. Average Monthly Salary by Educational Attainment

Educational Attainment	Average monthly salary (R$)
Complete tertiary	2,661
Incomplete tertiary	1,451
Complete upper secondary	925
Incomplete upper secondary	676
Complete lower secondary	689
Incomplete lower secondary	627
Complete primary	622
Incomplete primary	553
Illiterate	419

Source: Relação Anual de Informações Sociais (RAIS) 2004, http://www.mte.gov.br/.

economy has indeed exacerbated the velocity of change and posed new challenges to workers and firms.

With the surfacing in the 1960s and 1970s of a new economy—the knowledge economy—in which economic growth is as much a process of knowledge accumulation as of capital accumulation, the characteristics of and demands for employment shifted rapidly. A quickly globalized market in which trade became a key for economic growth also affected the nature of jobs. In countries that were inserting themselves into this economy, the skills demanded for jobs changed speedily. Moreover, a premium was placed on employees who had adaptable skills, could learn quickly, could communicate well, and could work in teams. Figure 6.1 highlights how these skill sets were swiftly changing in the United States between the late 1960s and the late 1990s. The changes are evident, with nonroutine, systemic, and analytical tasks growing at the expense of manual, routine tasks.

This rapid change in the nature of work in productive firms poses the question of how the education and training system is responding to this market reality. Are graduates and trainees exposed to a curriculum that adequately prepares them to meet the challenges posed by the work environment? Is the education sector responding accurately to the skill needs of the job market? Once again, the experiences of other countries are informative. Research done in the Arab Republic of Egypt some years ago demonstrated that Egyptian schools were good at teaching rote knowledge, facts, and rules, and very weak at teaching critical and independent thinking. A simple analysis suggests that the education system needed to be reformed to teach more critical thinking. However, further analysis determined that the public sector—which, like Brazil's, had better benefits, wages, and job security than the private sector—in fact demanded the rote knowledge. The education system was teaching the

Figure 6.1. Changes in Job Task-Skill Demands in the United States, 1960–98

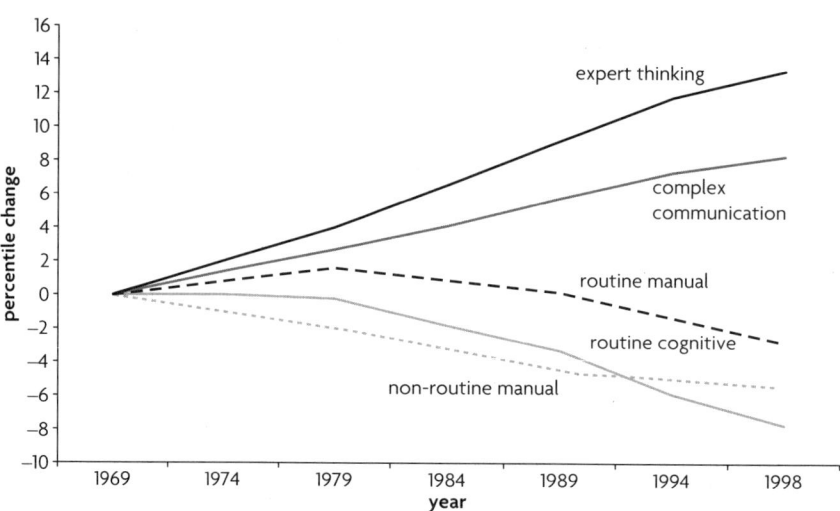

Source: Autor et al. 2003.

very skills that were best rewarded in the labor market. While this could pose an interesting question about what skills should be taught in Brazil given its job market, this chapter later presents evidence indicating that most Brazilian students are in an education system that teaches them neither rote knowledge nor critical thinking skills.

The political economy of industrial behavior further complicates the relationship between the education system and the job market. Consider the evidence of industrial behavior in Brazil documented by Tendler (2002). When a firm or production entity's competitiveness is founded on low-wage labor carrying out basic tasks, the external demand on public entities for more or better education may be limited. In fact, government policy may be subject to "local capture" by forces whose interest lies in limiting the expansion of education and the taxes associated with public finance of educational programs. Training programs inside the firm that are geared toward specific production processes may be favored instead of general skills. The result is, potentially, a deepening of the low-wage, low-skills, low-productivity trap that is aided rather than relieved by political forces. This kind of explicitly structural analysis is frequently absent from policy discussions about the paths that countries like Brazil need to take. On this issue, a recent IPEA (2006) report states that international competitiveness cannot be founded simply on low wages. Innovative behavior requires a human capital component, especially if the country is to avoid the trap of competitiveness driven by low wages that generates few linkages with other productive sectors of the economy or fails to stimulate the use of new technologies. In other words, a sustainable development strategy does not rely only on minimal technological adaptation using low-skilled (and low-wage) workers in manufacturing. Dynamic innovation that creates new forms of production—and new products—is also critical.

There is a general recognition that more and better education is necessary to improve employability and earnings but is not sufficient by itself to do so. In particular, Brazil's employment rates worsened for all workers during the 1990s, from those with no education to those with primary, secondary, and tertiary education (Berg, Ernst, and Auer 2006), suggesting, among other explanations, a mismatch of skills. This is precisely why improving and adapting workforce skills is crucial in a competitive global economy. In the Latin American region, Brazil lags considerably behind other countries such as Argentina and Mexico in the percentage of its population with more than six years of education, while a substantial percentage of Brazilian students have low reading skills (Berg, Ernst, and Auer 2006).

The Demographic Window: Heightened Urgency for Improved Skills

A final consideration is the importance of dealing promptly with the challenge posed by insufficient skills. Brazil is experiencing a temporary "demographic window" that must be rapidly exploited to promote rapid growth. Our analysis shows that Brazil is entering a 20-year period in which the economically active population is at its peak, with a decrease in the younger-than-15 dependent population (due to lower population growth), and with a still comparatively small population over 64. This suggests that generating labor income to sustain social policies and old-age pensions is within reach if employment creation is sufficient. Because of the large stock, productivity can peak during this window, provided that the working population is adequately skilled to perform at work. (The analysis for this can be found in appendix F.)

After this window closes, an increasing population over 64 and a shrinking labor force intensify the pressure, but this can be negotiated, as countries such as Finland and Norway have shown, through a highly skilled and productive economically active population.

The Formation of Human Capital in Perspective: Brazil and Its Competitors

When faced with the new realities that the knowledge economy was imposing on the workforce, countries reacted differently. The Republic of Korea and Singapore, followed by China a bit later, decided to invest heavily in basic education. These countries innovated mostly through the acquisition and adaptation of foreign knowledge and technology, in which basic skills—as discussed earlier—play a critical role. Other countries, like India, invested heavily in tertiary education to energize the creation and commercialization of knowledge, focusing particularly on information and communication technology development. When the millennium bug emerged, India was able to capitalize on the advanced skills that part of its workforce had developed, and it became a leader in ICT development. How did these countries improve their education systems to respond to the new realities?

The Republic of Korea

Korea took a sequential approach to expanding educational access. In the 1950s, access to elementary education was expanded, with a focus on producing a labor force that met the needs of an economy based on labor-intensive products and light manufacturing. Gradual provision of free compulsory education ensued, with an emphasis on cost efficiency. Such measures included double-shifts for classrooms, use of private schools to accommodate more students, and increases in classroom size. Heavy investments in basic education promoted quality and guaranteed access to all.

The 1960s brought a new focus on secondary education and technical and vocational education and training (TVET), which enabled the country to shift to capital-intensive and heavy-chemical industry. To accommodate the growing number of secondary pupils, private schools were used to absorb the new students, and classroom size was increased from 60 to 70.

The shift to emphasis on tertiary education since the 1980s then produced an economy based on electronics, high-tech, and knowledge innovations. The effort was initiated by the July 30 Educational Reform (1980), which broadened access to tertiary education by expanding the admission quota of colleges and universities and replacing university-administered entrance tests with a national-level examination. This effort was complemented by implementation of the Brain Korea 21 (BK21) policy in 1999, targeting what the Korean government considers to be the seven most important fields in science and technology for enhancing competitiveness in the global economy. The policy has several purposes: to develop world-class research universities, foster the creation of human resources through graduate schools, nurture quality regional universities, strengthen industry-university ties, and reform higher education in general. To accomplish this agenda, the government invested approximately US$1.2 billion into higher education over seven years. To date, increases have occurred in publications per participating professor in science and technology as well as in the humanities, in international patents and merchandising research, and in international exchange and collaboration.

Singapore

Singapore's experience was not much different. Singapore chose to use education as a major vehicle for overcoming its daunting post-independence challenges. Forging a close link between education and economic development was strongly emphasized in the nation-building process of this small city-state. In particular, developing a Singaporean identity through an integrated national education system was seen as key for economic survival.

The Five-Year Plan for 1961–65 was the first step toward boosting the educational standards of Singaporeans and uniting a disjointed educational system that included Chinese, Malay, Tamil, and English schools. The priority was to provide every child with at least six years of schooling. The plan consisted of six main educational reforms:

- Equal treatment for the four streams of education: Malay, Chinese, Tamil, and English
- Provision of a common syllabus for all school subjects in the four languages
- Compulsory bilingualism in all schools
- A common national examination system for the primary schools
- Universal free primary education
- Emphasis on the study of mathematics, science, and technical subjects.

By the late 1970s, social and economic indicators pointed to an increasingly rich and progressive Singapore amidst a cohort of developing countries still struggling to address national poverty. At the same time, Singapore's comparative advantage in labor-intensive manufactured goods clearly was being eroded as other Southeast Asian countries entered the global market. Singapore responded by attempting to transition to a more capital-intensive economy through a "Second Industrial Revolution."

The Singaporean government also decided to focus on improving educational quality after the enrollment surge of the 1970s. The New Education System was introduced in 1979 and academic improvements were evident by the late 1980s. Indeed, Singapore seems to have transitioned successfully to a knowledge-based economy focused on innovation and creativity, with an education system that performs exceptionally well as measured by results of standardized international mathematics and science tests such as the TIMSS (Trends in International Mathematics and Science Study) of 1995 and 1999.

In 1997, Singapore took its education reform a step further by introducing the vision of "Thinking Schools, Learning Nation," launched by then Prime Minister Goh Chok Tong. The vision hinges on the premise that, devoid of natural resources, Singapore's future viability and wealth depends on the capacity of its people to learn and to go on learning throughout their lives (Bon and Gopinathan 2006). Singaporeans were encouraged to continually acquire new knowledge; learn new skills; gain higher levels of technological literacy; and develop a spirit of innovation, enterprise, and risk-taking without losing their moral bearings or their commitment to the community and nation (Gopinathan 1999).

China

China reacted a bit later than Korea and Singapore. Indeed, China's recent accomplishments are in sharp contrast with the state of its education system prior to the reforms that began in the late 1970s. For three decades after 1949, total national resources devoted to education were relatively low and were heavily dependent on government allocation. The education infrastructure was weak, teachers were poorly paid, and large numbers of school-age children had no access to education (Tsang 1996). Two policy reforms played a particularly important role in transforming China's education system: decentralization of education financing and a curricular focus on science and technology.

Since the early 1980s, the financing of primary and secondary education has undergone a fundamental structural change. The official government policy for financial reform of basic education promulgated by the Chinese Communist Party in 1985 has two major components: decentralization in educational administration and financing, and diversification in the mobilization of educational resources. Further legislation passed in 1986 required governments at all levels to increase total expenditures for the basic cycles—and at rates higher than overall revenue growth—and to boost per pupil spending. More reforms followed in 1993, making the nine-year basic education cycle compulsory and encouraging private citizens and groups to participate in school development.

More recently, technology has also played a critical role in expanding access to education and raising its quality. For example, a pilot distance education program that started with 78 higher education institutions and the Central Radio and TV University now has over 2,000 off-campus learning centers around China, offering 140 majors in 10 disciplines, with a total enrollment of approximately 1.4 million students (Ministry of Education 2005).

In addition, TVET was identified as one of China's strategic educational priorities by the State Council in November 2005. Substantial efforts have been made to scale up and modernize the system in the past decade. Major rules and regulations governing organization of the TVET system were introduced in the Labor Law of 1994, the Education Law of 1995, and the Vocational Education Law of 1996. Reform elements have included the following: increased focus on access and equity; decentralization of control from the center to local governments; diversification of learning opportunities by opening up to nonstate providers; diversification of financing, including user fees; modernization of curricula and teaching methods; and promotion of a more integrated training system.

India

Soon after becoming independent, India placed tertiary education and science and technology high on its agenda for economic development. To meet the needs of industrial development, the first Indian Institute of Technology (IIT) was established in 1951 at Kharagpur (West Bengal). Support was received from the United Nations Educational, Scientific, and Cultural Organization (UNESCO), based on the MIT model. With the Soviet Union's assistance through UNESCO, a second IIT was established at Bombay (now Mumbai) in 1958. IIT Madras (now Chennai) was established with assistance from Germany the following year, and IIT Kanpur with help from a consortium of U.S. universities. British industry and the government of the United Kingdom supported the establishment of IIT Delhi in 1961. In 1994, IIT Guwahati was established entirely through indigenous efforts. In 2001, the University of Roorkee (first established as a college in 1847) became the seventh institution to enter the IIT family.

While taking advantage of experience and best practices in industrial countries, India ensured that the "institutions represented India's urges and India's future in the making" as Prime Minister Nehru said in 1956. The Indian Parliament designated them as "Institutes of National Importance"—publicly funded learning centers enjoying maximum academic and managerial freedom. The institutes offer relevant, high-quality programs in engineering, technology, applied sciences, and management at the undergraduate, masters, and doctoral level. Each offers its own degrees. To keep their Indian character, with equal opportunities for all, the IITs were designed to be fully residential for all students and most faculty members. This has permitted extensive student-faculty interaction beyond the classroom and optimal use of facilities. Most faculty and postgraduate students are involved in research and extension services. Admission is strictly according to merit through a highly competitive common entrance test.

Today, the IITs attract the best students interested in engineering and applied-science careers. IIT alumnae are well represented at the highest levels of responsibility in education, research, business, and innovation around the world. In 2005, the *Times Higher Education Supplement* ranked the IITs as the third-best engineering school globally after MIT and the University of California, Berkeley.

The key strength of the IITs has been their success in turning the best students into "creative engineers" and "engineering entrepreneurs." Initially, the IITs were criticized for contributing to the brain drain because some 40 percent of their graduates emigrated. However, the opening and fast growth of the Indian economy has transformed this weakness into a major strength for international cooperation and investment. Much of the success of Bangalore, for instance, is attributable to the phenomenon of reverse brain drain.

Obviously, the education systems of these countries still face numerous issues of equity and relevance. However, concerted government efforts in education clearly have been at the forefront of these Asian countries' success in the knowledge economy. While these reforms were occurring in the East, Brazil was slow to make educational progress. It was only in the 1990s that Brazil strove to make primary schooling universal—a very late start! Even today, Brazil is struggling to enhance quality in primary education and to make secondary educational enrollment universal. The next sections present a general overview of the challenges facing Brazil's education system.

Brazil's Education System and Its Readiness to Produce Human Capital for Growth

Clearly from the international experiences discussed in the previous section, nations focused on succeeding in the knowledge economy have undertaken broad and coordinated reforms of their education sectors as a key policy. Of course, as this study has emphasized throughout, enhancing the development of basic and advanced skills among the population is insufficient and must be complemented by policies that encourage investment by private firms in innovation and policies that maintain a sound enabling environment for business. This section focuses on the specific issue of human capital, which appears to be a critical bottleneck to Brazil's entry into the knowledge economy. A set of six tables (tables 6.3 through 6.8) summarizes the key issues that define Brazil's readiness to produce human capital for innovation and growth. The tables provide a snapshot of Brazil's primary, secondary, and tertiary education systems, as well as out-of-school opportunities for advanced training. To contextualize the challenges facing Brazil, international comparisons are shown throughout in italics. Each of the following tables is followed by key messages related to innovation readiness.

More detailed information on the individual components of Brazil's national education system is provided in appendix D (The Primary and Second Education System) and appendix E (The Tertiary Education System and Advanced Out-of-School Training).

Table 6.3. The Primary School System: Readiness for Innovation-Led Growth

Characteristics	Suggestive indictors (international comparisons in *italics*)	Implications for innovation-led growth
Access and coverage	Primary enrollment is "near universal" (98% in 2007), following 15 years of sustained effort. (*Brazil exceeds the Latin American average of 95%.*)	The primary system is still oriented toward expansion of coverage. It now needs to refocus on quality education for the emerging knowledge economy.
Rates of grade repetition and dropout	First-grade repetition rate is 28% (among the highest in the world). *Argentina's is 10%; Chile's, 1%; India's, 4%; Mozambique's, 26%; and the Philippines', 5%.*	Brazil's inordinately high grade repetition rates are understandably linked to the recent expansion of primary coverage; however, high repetition leads to age-distorted learning environments, which generally lead to early dropouts.
Cost of grade repetition	Annual cost of grade repetition in Brazil's primary and middle school budgets is US$600 million.	Excessive grade repetition not only consumes significant resources, it leads to age/grade distortions that undermine secondary-school quality.
The "typical" primary learning environment	In general, primary classrooms emphasize rote memorization, group recitation, and "correct" answers, rather than conceptual understanding and solution-oriented reasoning (Carnoy, Gove, Marshall 2007).	Current classroom pedagogy lacks the dynamic, interactive quality that goes beyond reading and arithmetic to prepare children with analytical skills and the capacity for innovative thinking later in life.
Science and math achievement	In 2003, Brazil ranked last in math and second to last in science out of 40 participating countries. *Brazil, 40th (math), 39th (science); Korea, 3rd, 4th; Mexico, 37th, 37th; Russian Federation, 29th, 24th; United States, 28th, 22nd.*	Brazil's exceptionally weak science and math achievement probably does not affect an elite minority of privately educated future scientists who will be educationally prepared to create new knowledge; however, it greatly affects the overall national capacity to use, adopt, and benefit from acquired technology.
Assessing educational quality and student performance	In 2005, the Ministry of Education administered the Prova Brasil, a US$25 million learning assessment of 3.3 million basic education students in over 42,000 schools.	The Prova Brasil provides a good basis for broadening and building upon a performance-based culture.
Standards	Standards are lacking in both learning performance and school operations. Many schools (especially in rural areas of the poorest regions) still lack suitable classrooms, basic furniture, and teaching materials.	Schools and municipal secretariats are adrift in goal setting, lacking standards to rationalize performance expectations and budgetary allocation.

(continued)

111

Table 6.3. (*continued*)

Characteristics	Suggestive indictors (international comparisons in *italics*)	Implications for innovation-led growth
Computers in the classroom	Computers are relatively rare or underutilized in primary instruction. Existing computers tend to be used by teachers and administrators. Brazil has 2 school computers per 100 students. *Korea has 28 school computers per 100 students.*	Computers are essential for preparing technologically literate graduates. Introduction in the classroom is even more important because medium- and low-income families cannot otherwise afford personal computers in their homes.
Preschooling leading to the primary grades	Lack of preschooling exacerbates the equity gap before young people even reach schools. Entering first graders from poor families are estimated to generally know about 400 words, compared with about 4,000 words by first graders from the top economic quintile.	Investment in early childhood education is needed not only to better prepare preschoolers for success, but to ensure that disparities in social equity are not reinforced from the outset.

Source: Author's compilation.

Key messages on primary education

- Educational policy makers have (rightfully) focused on extension of primary coverage, literacy, and educational equity for the past 15 years.

- Brazil's primary schools are deficient in teaching the basic reading, math, computer, and science skills that are the foundation for broad social participation in the knowledge economy.

- Brazil's primary schools are highly deficient in laying foundations for conceptual reasoning, solution-oriented thinking, and the scientific method—"ways of thinking" that lead to a flexible, competitive, and productive national workforce.

- The federal government needs to spearhead reform to energize the national curriculum, establish minimum operational standards for schools, and encourage accountability based on performance. The recent Plan for Educational Development (PDE) contains precisely the rules of the game for this new output-based incentive program for states and municipalities.

- Workers with sound basic skills who can use and adapt to new technologies are needed on the shop floor; and economically at least, these workers are no less essential than the engineers and managers who introduce new technologies and set the pace for productivity growth, nor the public and private sector researchers whose R&D may lead to new discoveries and applications.

Table 6.4. The Secondary School System: Readiness for Innovation-Led Growth

Characteristics	Suggestive indictors (international comparison in *italics*)	Implications for innovation-led growth
Access and coverage	Gross secondary enrollment has improved dramatically—from 15% in 1990 to 76% in 2004. *OECD average is 92%, Chile is at 80%, Finland at 97%, and Korea is at 89%.*	With plateaus in sight for enrollment, Brazil is in a strong position to expand human capital by reorienting its poorly performing secondary system toward innovation and competitiveness.
Educational attainment for population of postsecondary age	Average educational attainment for population 15 years and older is still only 4.3 years. *Argentina is 8.8 years, China is 6.2 years, Korea is 10.5 years, Mexico is 7.2 years.*	Despite the significant advances in secondary enrollment, much work remains on raising completion rates and providing the most basic levels of literacy and numeracy training.
Secondary dropout and completion rates	Secondary school dropout rates remain unusually high, and secondary school completion rates remain unusually low. This reflects educational supply deficiencies, especially in rural areas.	The key to success at the secondary level is to improve quality. Efficiency gains at the primary level (in part through a reduced retention rate, which costs an estimated US$600 million annually) could significantly help to improve secondary-school quality. In the longer term, improved completion would more than pay for itself through a more productive labor force.
Impact of high retention	Because of the high retention rates in primary grades, secondary schools contain many older students with extremely weak skills. Their situation is worsened by a standardized curriculum socially geared toward younger children. School dropout tends to be deferred to secondary education rather than avoided.	High primary retention complicates secondary schooling through age/grade distortion. The older evening students could be placed on an accelerated basic-skills curriculum similar to EJA (*Educação de Jovens e Adultos*), in which all students also receive instruction in workplace skills such as communications, computer use, and negotiation.
Reading and language achievement	About half of Brazilian 15-year-olds have difficulty reading or cannot read (international PISA [OECD Programme for International Student Assessment] test). Only 9% of 8th graders are performing at a satisfactory level in Portuguese (SAEB). *Only 6% of Korean 15-year-olds have difficulties reading or cannot read (PISA international test).*	A weakly literate workforce imposes costs and foregoes benefits at every level of the economy's productive processes.

(continued)

Table 6.4. (*continued*)

Characteristics	Suggestive indictors (international comparison in *italics*)	Implications for innovation-led growth
Math and science preparedness	More than three-fourths of Brazilian 15-year-olds cannot perform basic math or have significant difficulties doing so (PISA); only 7% of 11th graders perform at a satisfactory level in math (SAEB) *Math scores are below Mexico's and Indonesia's, and far below "high scorers" such as Korea. (Relatively, Brazil's science scores are similarly low.)*	Math, science, and technological literacy are essential—not only to produce scientists and engineers but to create a workforce able to use, adapt, and disseminate new ideas and technology. For economic success, Brazil will need to make major compensatory investments to improve math and science performance at the secondary-school level.
Instructional hours per week	Average hours in the classroom per week are 19.1. *In Mexico, it is 25 hours; in Korea it is 30.3.*	The number of instructional hours for academic courses and vocational training needs to be increased, especially if nonacademic curricula such as civic training, sex education, drug prevention programs, and so forth are to be maintained.
Social equity in secondary education	The poor are significantly less likely to complete secondary education. The completion rate for children from families in the highest decile of socioeconomic status (SES) is over 90%. The completion rate for children from the lowest decile of SES is 4%.	Efforts are needed to keep poor children in school longer—for example, conditional cash transfers (CCTs) as incentives to secondary attendance and savings accounts to attract students and keep them in school. Completion rates will increase if poor families perceive that secondary education produces marketable job skills as well as entry to higher education.
"Nonacademic" secondary tracks	The secondary curriculum is highly oriented toward preparation for university entry. Nonacademic students tend to be segregated as night students; however, their curriculum is university oriented and training is scant in technological fields. Intellectually capable students older than 20 who do not possess diplomas have few opportunities to receive advanced skills training.	A validation exam that can be used as an equivalent secondary credential exists but is not widely used. The validation exam should be readily available, preferably online, and geared toward providing technological training opportunities for persons older than 20.

Academic and workforce tracking	Virtually all students are on an academic track with a pre-university curriculum even though the vast majority will not attend university. (55% attend evening school.)	Secondary schools need to provide options for students to pursue nonacademic as well as academic training.
Vocational training	Vocational training in secondary schools is rarely offered. Indeed, Brazilian legislation moved vocational training to postsecondary education. For nonacademic students who have achieved mastery of basic skills, vocational training opportunities could be offered and welcomed.	Secondary schools need to provide nonacademic students (especially older night students) with a greater range of training choices, including channeling them to short, focused postsecondary courses or to training and education offered through a broad support network for industrial workers known as the S-system.
Retention and dropout	Schools contain many older students who have been retained in lower grades because of weak performance; however, the curriculum is geared toward younger children, tending to defer rather than solve the problem of dropping out.	The older evening students could be placed on an accelerated basic-skills curriculum similar to EJA, in which all students also receive instruction in workplace skills such as communications, computer use, and negotiation.

Source: Author's compilation.

Key messages on secondary education

- Progress in secondary education depends on more than additional financing to expand the number of secondary-age young people in school.

- Functional reading abilities, math skills, and technological literacy need to be improved across the board at the secondary level.

- New academic tracks are needed to prepare secondary students for new jobs in the knowledge economy that do not necessarily require university education.

- Far more attention must be paid to school-to-work transitions.

- Primary-education quality is the key to quality in secondary education; just as secondary-education quality is the key to quality in tertiary education.

Table 6.5. Features Related to School Performance and Governance

General features related to performance and governance	Suggestive indictors (international comparison in *italics*)	Implications for innovation-led growth
Size and scale of the educational system	Primary (1st–8th grades) enrollment is 45.1 million students. Upper-secondary enrollment is 9 million students. *In China the figures are 188.5 and 31.2 million; India, 185 and 35 million; Mexico, 21.7 and 3.4 million; United Kingdom, 7 and 3.3 million; and the United States, 37.6 and 11 million.*	The vast size of the public education system and recent achievement of high coverage provides an unprecedented opportunity to shape the nation's future through education for innovation.
Role of federal government	The federal government sets policy and provides budget funds. It does not deliver services.	Through its policy setting and budget allocation roles, the federal government has vast scope to set performance standards, reshape curriculum content, finance pilot initiatives, and broadly encourage innovation.
Comparative advantage of municipalities, states, and private sector (versus the federal government)	Municipalities employ about 48% of all teachers; states employ 39%; private sector employs 12%. These are the implementers who convert "reform" into reality.	Municipalities, states, and private schools must be provided with resources and support to reshape the formation of human capital for an innovation-based economy.
Expenditure on education	National public spending on education increased from 3.9% of GDP in 1995 to 4.3% of GDP in 2005. *This is about average for Latin America and the Caribbean. The OECD average is 5.5%; China is 2.1%, Korea is 4.2%, Japan is 3.6%, Mexico is 0.2%, the United States is 5.5% (data from 2002).*	While "more" expenditure would help, the more difficult questions involve the priorities and expectations for and the distribution and efficiency of educational expenditure.
School principals	More than 60% of Brazilian principals obtain their jobs based on political criteria.	A professionalized certification process is needed to ensure that every principal understands the learning process and is administratively competent to manage a school.

| Community participation | Brazil has a long history of community participation in schools. *Brazil enjoys an advantage in this area that many other countries might envy.* | Workforce-oriented education could be energized by bringing in more speakers from the community, organizing open school events, emphasizing internships and apprenticeships in local enterprises, and fostering better understanding of changing labor markets by forming school-to-work partnerships with local industries and firms. |
| Accountability based on measurable performance standards | Budgetary allocation is based on standardized formulae with few incentives to recognize or reward high performance. | The incipient "culture of evaluation" must be preserved and deepened while at the same time avoiding too much testing and redundant testing by multiple levels of government. |

Source: Author's compilation.

Key messages on school performance and governance

- All levels of the education system need to be functionally accountable for educational performance.

- Students' capacity to innovate is not the only criteria against which educational performance should be assessed; however, it needs to measurable and assessed.

- Higher standards and accountability for educational performance need to be institutionalized at all levels of the public educational system.

Table 6.6. Teachers and Teaching in the Primary and Secondary Schools

Characteristics	Suggestive indictors (international comparisons in *italics*)	Implications for innovation-led growth
The number of teachers and pupil-to-teacher ratios	Brazil employs approximately 1.5 million teachers. The student-teacher ratio is 22.4 in the primary system and 17.5 in the secondary system. *Argentina's ratios are 19.1 and 19.8, respectively; Chile's, 33.9 and 32.7; China's, 21.9 and 18.8; India's, 40.2 and 32.5; the United States', 15.5 and 15.5; and the OECD averages are 16.5 and 13.6.*	Brazil has struggled to expand enrollment without dramatically increasing the pupil-to-teacher ratio. Major efforts have been made to train and hire teachers and to reduce regional and social inequalities. To ensure quality education with a greater share of math, science, and technical content, issues of teacher quality and pupil-to-teacher ratios will need to be examined.
Salaries	Brazil's highly unionized teachers currently earn 56% more than the national average salary. *In OECD countries, teachers earn 15% less than the national average.*	Teacher salaries may be too high in general and too low for high-performers and those with specialized skills. Higher entry salaries may be needed to attract qualified teachers, with fewer automatic pay raises for seniority. Higher-education grants may be needed to equip teachers with knowledge-economy skills.
Teacher salaries as share of total educational budget	Teacher salaries currently absorb 75% of total national educational expenditure. *The ratio of teacher salaries (primary + secondary) to total expenditures is among the highest in the world and three times the OECD average.*	High fixed salary costs will lead to intense national debate as reforms for higher quality seek to expand other categories of expenditure.
How teachers teach	In pegging Brazil's "typical" teaching environment to international standards, a recent qualitative assessment of 3rd-grade math classes reported as follows: "… More time copying lessons and instructions from chalkboard… heavy reliance on whole-class recitation… individual rather than group-oriented work … high degree of talking, horseplay, and inattention in classrooms … teachers check that students 'do the work' rather than assess competence … teachers ask few questions while teaching, almost none requiring conceptual or analytic responses."	Brazilian teachers tend to be trained in the philosophy rather than the implementation of teaching. Their emphasis on rote memorization and recitation needs to be replaced with pedagogy based on active learning, the scientific method, and capacity to think outside the box.

Teacher training	Some 34% of teachers have received no university education (disproportionately in poor communities and rural areas). Only 21% of teachers hold master's degrees. As in most of the world, math and science teachers are in noticeably short supply. *Zero percent of Korean teachers have received no university education, and 93% of Korean teachers hold master's degrees.*	Certification of unlicensed teachers—for example, through distance learning—needs to be extended. Teachers need to be trained and retrained in teaching methods that are more active and learner oriented.
Short training courses	Short training courses are available, but they seldom focus on student learning. Attendance in these courses is generally a poor predictor of improved classroom instruction.	Promotions and career advancement should be linked to performance rather than to attendance at training courses or seniority. Short courses need to produce better teachers. Teacher capacities in math, science, and technology need to be upgraded.
Absenteeism	High teacher absenteeism is endemic, especially in rural areas and the poorest schools.	Continuous teacher absenteeism is costly in both financial resources and educational quality. Rewards for better performance can be allocated individually to teachers or collectively to schools.

Source: Author's compilation.

Key messages on teachers and teaching in the primary and secondary schools

- Teachers need more training in math, science, and technology.

- Teachers need more training in student-oriented teaching methods that prepare children to think conceptually, exercise creativity, and ask questions.

- Teaching salaries should be high; but the salary costs cannot be allowed to crowd out other options for educational improvement.

- Good teaching needs to be rewarded with tangible incentives.

Table 6.7. Advanced Skills Training Outside the Schooling System

Characteristics	Current status	Implications for innovation-led growth
Coverage	An extensive network offers 2,300 vocational courses per year. The annual enrollment of roughly 15.4 million students is the largest system in Latin America.	Training and retraining opportunities for young and mature adults are needed to improve or upgrade their industrial and commercial skills.
Linkages between training and private firms	The S-system, developed over 50 years, comprises 9 training and technical assistance networks, operating in partnership with employers.	Close linkages increase probability that training offered is relevant to firm needs.
Content and competencies	Much training is still based on the traditional approach of "Taylorism." Programs are not competency based. Most are not up-to-date.	Current training content and competencies may be relevant for many traditional firms, but not for adoption of innovation in a knowledge-driven economy.
Financial sustainability	The training system is 85% financed through a compulsory 2.5% payroll tax on private companies, with the other 15% coming through contracts with the public sector.	The financial sustainability of the system needs to be analyzed and ensured over the long term.
Advanced on-the-job training	Only the most innovative firms provide advanced on-the-job training.	A virtuous circle exists among firms that are already innovative because they tend to invest more in continually updating the skills of their employees.
On-the-job training for lower-skilled employees	Firms report significant on-the-job training, primarily to provide workers with basic skills not provided by schools but not to provide higher-productivity skills.	If the schooling system would indeed provide all graduates with basic and advanced skills, firm training content could be better focused on skills oriented to more efficient and effective operations and, thus, higher productivity.
Distance learning	Employer surveys indicate unsatisfied demand for distance learning usable for in-firm training (professional skills and employee attitudes).	A potentially significant resource is not being well utilized.
Unemployed youth	A recent survey of the unemployed indicates significant lack of access to S-system courses.	There is a strong unmet need for outreach and training programs geared toward youth.

| Older workers | Older, out-of-school workers have few opportunities to gain new high-tech skills, receive a secondary diploma after the age of 20, or validate on-the-job learning through a career-enhancing credential. | Validation exams should be readily available, preferably online. They should be geared toward persons beyond normal school age who wish to obtain a secondary diploma or demonstrate competencies and knowledge. |
| Linkages between informal VET (vocational education and training) programs and formal education | Contrary to the Education Law, the national system does not formally validate skills and competencies acquired outside the formal education system in VET courses. | Further training opportunities are needed for older, out-of-school workers as well as increased formal validation to act as incentives for acquisition of out-of-school skills and competencies in VET. |

Source: Author's compilation.

Key messages on advanced skills training outside the schooling system

- A 2.5 percent payroll tax provides a sustainable financial basis upon which to operate, modernize, and expand Brazil's extensive S-system for vocation training.

- Existing vocational education programs are mostly geared to the needs of traditional firms. They need to be made relevant and productive in serving the needs of innovation-ready companies.

- Outside the regular S-system, few training opportunities are currently available for unemployed youths, thereby further marginalizing this population and losing their potential productive contributions.

- Internet-based learning for advanced skills training is significantly underutilized as a resource for out-of-school technical training.

- Many Brazilian firms are forced to provide basic-skills training that should have been provided by the national school system. This represents a waste of both resources and opportunity. Schools must provide graduates who are ready to learn and innovate when they enter the workforce.

- Strong linkages should be built between secondary schools serving older students and employers, technical and vocational service providers, and the S-system. However, technical and vocational training should be left to postsecondary education, where it should be provided primarily through short, flexible, narrowly focused courses.

- A validation exam already exists in Brazil; however, it is not widely used. The validation exam should be geared toward adults, underscoring the idea that learning is not confined to schools but is a lifelong enterprise that proceeds through successive phases of training.

Table 6.8. Tertiary Education and Readiness for Innovation-Led Growth

Characteristics	Suggestive indicators (international comparisons in *italics*)	Implications for innovation-led growth
Access and coverage	Only about a quarter of young adults ages 18 to 24 are enrolled in tertiary education institutions. *Argentina enrolls 64% and Chile enrolls 47%. The average for Latin America and the Caribbean is 30%.*	This results in a small proportion of the labor force with the high-order skills needed to understand, adapt, improve upon, and disseminate new knowledge and innovation.
Equity	A very low proportion of students come from low-income families. At UNICAMP (Universidade Estadual de Campinas), 10% are low-income students compared to their 69% proportion of the overall cohort in Brazil.	There is loss of talent (young people who could have contributed to application or generation of innovations).
Overall quality	The system is very heterogeneous, with some "islands of excellence" (i.e., a few very good public and private universities) surrounded by many poor-to-average-quality institutions.	In most cases, graduates do not have the high-order skills needed by the knowledge economy.
World-class universities	No Brazilian university consistently ranks among the world's top 100 universities. *The best universities in China, India, and Russia rank higher than the best universities in Brazil.*	Most Brazilian universities are unable to produce the graduates and research needed to fuel innovation in the economy.
Access to top universities	Access to top universities is highly competitive. Ratio of applicants to places has grown in top universities, e.g., 16 to 1 in UNICAMP.	Graduates from elite universities are likely to be more competitive in the labor market.
Assessment of secondary-school graduates	Introduction of the ENEM (Exame Nacional do Ensino Médio) aptitude test at the end of secondary education has leveled the playing field in admission of students from less-privileged backgrounds. Proportion of high school graduates participating in the exam rose from 7% to 82% within seven years.	The ENEM is likely to enlarge the pool of graduates from low-income backgrounds.
Private tertiary education institutions	Many private tertiary education institutions do not screen applicants. Ratio of candidates went down from 3.4 to 1 in 1980 to 1.4 to 1 on average in 2004. Proportion of full-time instructors is only 20% vs. 83% in federal universities.	Graduates from low-quality private institutions are unlikely to contribute positively to the knowledge economy.

Faculty qualifications and productivity	Despite significant progress, academic qualifications are still insufficient. Proportion of academics with a doctoral degree rose overall from 15% in 1994 to 21% in 2004 (and doubled from 21% to 42% at federal universities). Teaching staff are civil servants; tenure is a right, based on seniority rather than performance.	Professors and researchers are still insufficiently qualified, and there are few incentives to evaluate and reward teaching and research productivity.
Assessment and performance standards for universities	The government has instituted a comprehensive quality assurance system, SINAES (Sistema Nacional de Avaliação da Educação Superior), including assessment of learning outcomes ENADE (Exame Nacional de Desempenho dos Estudantes), institutional self-assessments, and external evaluations. Results indicate gradual improvement on average, though extreme unevenness and no penalties for low quality.	Universities can only train students to the quality of their own standards.
Internationalization	There is very little international mobility of students and faculty. In 2005, only 2,075 students were officially sponsored for graduate studies outside Brazil (2% of total postgraduate student population). Only 1,246 foreign students are enrolled in Brazilian universities.	The outlook of graduates is internal to Brazil, making it difficult for them to compete in a global economy.
Production of specialized technical skills for the labor market	A disproportionate share of students comes out of the social sciences and humanities. Only 19% of students major in science and engineering. *In Chile, 33% of students are enrolled in science and engineering; in China, it is 53%.*	There is a lack of graduates with appropriate professional skills for innovation-oriented firms.
Production of middle-level professionals and technicians	Only 2% of the student population is in nonuniversity institutions or short-duration professional programs.	There is a deficit of qualified technicians and middle-level professionals.
Lifelong learning opportunities	There is lack of articulation among the S-system, nonuniversity professional programs, and university programs. Too few pathways (mutual recognition of equivalences) facilitate student mobility among different types of institutions.	There is a lack of opportunities for skills upgrading, which is highly needed in sectors and firms undergoing innovation-induced productivity changes.
Employment rates of university graduates	Unemployment is rising among graduates. The proportion of unemployed university graduates is 16.4%, compared with the overall national unemployment rate of 9.3%.	Universities are not training students with skills actually in high demand.
University-industry linkages	A culture of collaboration with industry is lacking.	Most universities are unlikely to support local firms or contribute to regional development.

(continued)

Table 6.8. (*continued*)

Characteristics	Suggestive indicators (international comparisons in *italics*)	Implications for innovation-led growth
Governance and autonomy	University governance suffers from excessive central control.	Universities are unlikely to become more innovative and responsive.
Resource allocation	Performance-linked budget allocation mechanisms are absent.	Universities have no incentives to become more innovative and responsive.
Resource utilization	Federal universities have excessive unit costs.	Resources are diverted from supporting expansion and quality-improvement goals.

Source: Author's compilation.

Key messages on tertiary education

- With only one-quarter of the relevant population group attending a tertiary education institution, Brazil has the next-to-lowest gross enrollment rate among the largest Latin American countries. Only 8 percent of the labor force holds tertiary-level qualifications.

- Access to tertiary education, especially the most prestigious universities, is heavily skewed against students from low-income families.

- Unlike China, India, and Russia, Brazil has no university ranked among the top 100 in the world.

- Research is concentrated in a very small group of elite public universities. The second tier of universities (public and private) has some pockets of research strength; however, most universities conduct little, no, or very low-quality research (usually in nonscientific, nontechnological fields).

- Few universities collaborate meaningfully with the productive sectors.

- There are too few students in science and engineering programs, as well as in nonuniversity technical institutions and short-duration professional programs.

- Overall, the quality of research and teaching has been gradually improving. Brazil has pioneered assessment tests to measure student learning in conjunction with external institutional evaluations.

- There is a gap between skills taught and labor force demands: Graduate unemployment has been rising faster than overall unemployment, while employers indicate that they cannot always find qualified personnel.

- The present centralized governance system significantly constrains university performance.

- With limited use of performance-based budgeting, public universities have few managerial or financial incentives to use resources efficiently or respond to labor market and social needs. In particular, costs at federal universities are excessively high.

Conclusions: Schooling and Beyond

This chapter has focused on issues of educational attainment and quality insofar as they are keys to generating human capital that can support innovation-led growth. We conclude this review with a discussion of how learned skills are used, or in other words, the overall relevance of the educational experience. It bears restating the guiding theme of this study: There are concerns that Brazil is inadequately prepared to compete in an increasingly globalized world and that, despite advances in addressing serious deficiencies in recent decades, the education system remains a weak link.

We have already raised questions about the relevance of education in Brazil based on internal efficiency measures such as repetition rates and SAEB scores. A large segment of the student population is clearly not performing to standards that meet the current requirements, much less the future requirements, of a competitive global economy. This problem has serious consequences for equity, which is discussed below, as well as for efforts to reduce poverty. But perceptions about the relevance of the educational experience in Brazil may also be affecting a more immediate outcome: school dropouts.

Ioschpe (2004) estimates that roughly 7 percent of primary school students drop out annually, while almost 8 percent drop out in middle school. Most discussions of school dropouts focus on socioeconomic factors, but there is a growing body of evidence linking features of schools—including learning—with the decision to remain enrolled (Hanushek and Lavy 1994; Bedi and Marshall 2002; Marshall 2003; Lloyd et al. 2003). In the Brazilian case, low-income students frequently fail grades and may feel increasingly uncomfortable in the classroom with younger students. Furthermore, if they are not learning anything, or view the curriculum as irrelevant to their lives, then perceptions of the importance of schooling are likely to be reduced— and dropout will occur.

Relevance is also an issue in school-to-work transitions. First, the failure to generate basic skills in the early grades has consequences for global competitiveness. The De Negri et al. 2006 sector study on work in Brazil addresses this point and situates deficiencies in Brazilian education more concretely within the context of skills deficiencies. Hanushek's recent (2007) review of school quality and development is another reminder that school quality is not necessary just to make better citizens or help workers earn a living wage—the education of the average worker has far-reaching consequences for all sectors of society.

In Brazil the current competitiveness profile is marked by poles of excellence in specific areas (Petrobrás, Embraer, Embrapa, and others). These are large enterprises that use internationally competitive technology and highly educated managers and technicians. But for a more microdevelopment strategy that harnesses innovation at the small and mid-size firm level, a steady supply of quality workers is required. This is the foundation for sustainable, broad-based economic growth that is critical for creating good jobs and fostering human development. Unfortunately, in an era of global competition, the current state of education in Brazil means it is likely to fall behind other developing economies in the search for new investment and economic growth opportunities. As a result, the overall size of the pie to be divided among its citizens will not be large enough to keep them on a sustainable, development-oriented growth path.

Finally there are school-to-school transitions that are also affected by quality and perceived relevance. Based solely on the quantity of education, it is easy to be optimistic about the equalizing potential of education, especially given the rapid increase in basic education completion rates and secondary school enrollments. For example, multivariate analyses of earnings show that primary school (grade 8) graduates earn 50 percent more than *analfabetos* (workers lacking literacy skills), while high school graduates (*Segundo grau*) earn almost another 50 percent more than primary graduates (data from the Labor Ministry's 2004 RAIS, Relação Anual de Informações Sociais). But these kinds of returns are not guaranteed in the future because more and more young people are obtaining these credentials, not just in Brazil but in other countries as well. This in turn highlights the school-quality premium and the ongoing challenges in Brazilian education to provide equal opportunities.

The expansion of schooling in Brazil in recent decades represents an important step in the process of building a more just society and competitive economy. But when we see the gap between actual learning and the skills that children need to be successful in later school-to-school and school-to-work transitions, the potential long-term dangers facing Brazil are very real. Other countries are also expanding access, but if they are doing a better job of improving basic skills—let alone more-advanced cognitive skills—then Brazil actually may be falling farther behind rather than catching up.

CHAPTER 7

How Brazil Can Foster Innovation

How to capture innovation for accelerated growth? This chapter addresses that question across the six critical areas discussed in the preceding chapters—developing an enabling environment for investment, creating and commercializing knowledge, acquiring foreign knowledge, leveraging and disseminating the use of technologies, improving basic education and skills, and expanding tertiary education. Taken together, the recommendations made in this chapter constitute the beginnings of a comprehensive national plan.

In the six sections that follow, our primary recommendations are underlined. Many of the recommendations here were discussed at the 2007 "Global Forum on Building STI Capacity for Sustainable Development and Poverty Reduction."[1] Lessons were highlighted, and many case studies were presented at the forum. However, for ideas such as these to be translated into action, a far more inclusive national process will be needed in Brazil. That process requires greatly increased public awareness of what's wrong, including the costs now and into the future of continued sluggishness in responding to the challenge. Ongoing analysis and a vigorous national debate can lead to a workable integrated national strategy.

The first section, below, focuses on the development of an enabling environment for private sector investment. Most of the items in this section are not new recommendations coming out of our analysis, but are dimensions of reform that already have been identified in previous World Bank work. However they are crucial to the implementation of the other recommendations made in our report and deserve, therefore, to be repeated in this context. The subsequent five sections focus more directly on the policy reforms and measures linked to our analysis of the determinants of innovation in Brazil.

Developing an Enabling Environment for Private Sector Investment

Stay the course in continuing to improve the basic macroeconomic environment

Over the past decade, Brazil made substantial inroads in one area of the enabling environment in particular—macroeconomic stability. Fiscal policy contributed to a reduced public-debt-to-GDP ratio and to increased sustainability of the public debt (mostly through increased tax revenues). Monetary policy based on an inflation-targeting framework and a flexible exchange-rate regime reduced inflation from 12.5 percent in 2002 to 3.1 percent in 2006, as well as reduced foreign-exchange risks. In addition, debt management has drastically reduced public external debt.

Facilitate firm-level investment

Despite a good mix of economic policies and a highly favorable external environment, economic growth has nevertheless remained under 3 percent. The rapid growth that has transformed countries elsewhere has continued to elude Brazil.

Why? One significant obstacle discussed throughout this report (in particular, in chapter 5) is that the overall environment continues to pose significant obstacles for firm-level investment. Firms' correspondingly low rates of investment in R&D are central to their difficulties in increasing productivity through innovation. The analyses undertaken for this study along with consultations with the business sectors highlight the kinds of obstacles to R&D investment that remain. Recommendations for overcoming these obstacles are as follows:

- *Reduce the tax disincentives to R&D investment.* Overall taxes as a share of GDP were 37 percent in 2005, which is unusually high by international standards.

- *Lower the cost of capital.* According to the *World Development Indicators* (World Bank 2006d), Brazil's interest-rate spreads were 38 percent in 2005, which is highly unfavorable compared with competitors such as Russia (7 percent), Mexico (6 percent), China (3 percent), and Korea (2 percent).

- *Continue to open the economy to foreign competition.* Despite relative progress, the Brazilian economy remains fairly closed by international and regional standards. Trade amounted to just 37 percent of GDP, well below the average of 44 percent for the Latin American and Caribbean region.

- *Challenge the rigidity of Brazil's labor markets.* All in all, it is simply too difficult to hire and fire workers. A recent survey shows that the rigidity-of-employment index in Brazil is at 42.0, above the regional average of 37.1.

- *Reform the social security system.* A generous and inequitable social security system represents the lion's share of public social expenditures in Brazil. Recent attempts at reform have yielded minimal results and have not solved the major deficit behind this system.

- *Address infrastructural deficiencies that increase the costs of production and of doing business.* Logistical costs as a share of GDP, for example, are much higher in Brazil (24 percent) compared with Chile (16 percent) or Mexico (18.5 percent). Other elements of the institutional setting are similarly unfavorable and need to be reformed, namely excessive bureaucracy, corruption, and an inability to control violent crime. The costs and delays in contract enforcement are especially damaging for businesses.

Improving the investment climate so that the private sector can reach its investment potential is the single most important action the government could take in strengthening private investment in innovation. The most politically complicated areas for reform touch on labor, pensions, taxation, and trade. Previous World Bank reports have discussed policy options in these areas in detail.[2]

Significant action is particularly needed in three areas:

Improve the efficiency and intermediation of the banking system
Brazil's banking system is highly sophisticated, but its efficiency can be improved to reduce costs and pass savings along to consumers.

Increase the use of public-private partnerships to amplify leveraged investments in infrastructure
The Public-Private Partnership Law was approved but has had little practical impact. Its implementation is critical. Other opportunities for public-private collaboration can be further strengthened, such as the OSCIPs (*Organização da Sociedade Civil de Interesse Público*).

Improve governance
Several actions are required. These include (a) simplification of procedures for doing business in Brazil,[3] (b) more effective auditing and anticorruption mechanisms to reduce "leakage," (c) government streamlining to reduce size and administrative expenses, and (d) improvement of justice system effectiveness to prevent crime and violence and to improve the system's capacity to enforce contractual relationships. A detailed analysis of the significant bottlenecks in Brazil's judicial system can be found in a recent World Bank report (Hammergren 2004).

Creating and Commercializing Knowledge and Technology

As a share of GDP, Brazil's R&D expenditure is somewhat above average for its level of per capita income; however, when compared with the levels of expenditure in China and India—two of its most important BRIC comparators, both of which have much lower per capita incomes—its level of R&D expenditure is too low. Given that technological innovation is becoming increasingly crucial to international competitiveness—and that Brazil's main comparators are dramatically increasing their own R&D efforts—Brazil needs to significantly improve its effort to create and commercialize knowledge. The following key actions are required:

Increase private R&D

Brazil spends 1.1 percent of GDP on R&D, but mostly through government expenditure. The private sector accounts for only 30 percent of total R&D investment. Based on the experience of OECD countries, as well as firm-level studies in Brazil itself, this private R&D investment rate is far too low. Actions to improve the enabling environment—recommendations to further liberalize the trade regime and improve domestic competition policy, as described above—will offer strong incentives for firms to increase their R&D effort. A strengthened export orientation that places firms more squarely in the competitive international market will also serve to increase their R&D efforts. However, as discussed in chapter 5, R&D effort and export orientation are correlated with firm size. Larger firms are more able and likely to respond effectively to the increased competitive pressure. Therefore, government policy should focus on improving R&D effort by small and medium-size firms. Some of the possible actions include the following:

- *Expand the mix of public financial instruments that foster private R&D to include more risk-sharing, matching grants, equity instruments, and loans.* Currently, government support for R&D is primarily given through grants (primarily to university research) or as tax incentives to large firms that often would have undertaken the research anyway. Tax incentives are of little use to new start-ups with no profits to offset. Carefully designed instruments that provide risk-sharing mechanisms, matching grants, equity instruments, and loans may provide far greater benefit for small and medium-size firms.

- *Improve monitoring of how well different instruments and mechanisms work.* Currently, there is very little evaluation of the effectiveness of government mechanisms to encourage R&D by private firms. Evaluation results should be used to improve the programs and instruments, redeploying resources to those that are working well and closing down those that are not.

- *Improve cost effectiveness of fiscal incentives for R&D.* To date, fiscal incentives for private R&D have mostly benefited larger firms, many of which would have undertaken research with or without incentives. Thus, an effort should be made to improve the design of incentives so that they result in additional research and to include provisions that make incentives more relevant and attractive to new and smaller firms (for example, loss carryovers and simple procedures).

- *Improve operation of the sectoral science and technology (S&T) funds to provide greater flexibility across sectors and to increase interaction among academia, research institutes, and the private sector.* The sectoral funds represent a solid advance in greater investment resources for R&D. However, their operations are overly restrictive in focusing primarily on support for university research along disciplinary lines. Strictly earmarked compartmentalization of the funds by sector should be avoided.

- *Improve interaction among public labs, universities, and the productive sector.* Government support mechanisms should encourage cross-fertilization among academia, research institutes, and the private sector. This could be accomplished by making some government support contingent upon the participation of more than one key actor—as is done, for example, by many research-support programs in the European Community and the United States.

Improve public R&D

- *Increase public R&D resources.* As a share of GDP, Brazil spends far less on R&D than key lower-income competitors such as China and India, and far less still than the OECD average. Brazil needs to increase its public R&D effort—not only for universities and firms, but by the public sector itself.

- *Strengthen public R&D in key strategic areas, such as natural resources, renewable energy, biotechnology, and nanotechnology.* Currently, most public R&D is undertaken by sectoral ministries such as defense, agriculture, industry, health, and the environment. While this is appropriate, new areas related to natural resources, renewable energy, biotechnology, and nanotechnology hold significant future potential. Mechanisms are needed to allocate funds appropriately and coordinate new initiatives.

- *Increase support for R&D in universities.* Universities have become ever more important R&D agents worldwide. While many of Brazil's programs do support university research, the volume is still small by global standards. More resources should be channeled through competitive projects with built-in monitoring and evaluation. University research should be aimed not just at advancing the frontiers of science, but at knowledge that is relevant to Brazil's social and economic needs. Based on merit, funds should be available to researchers at private as well as public universities.

- *Improve monitoring and evaluation of public research, with the results used to reallocate funds according to performance.* There is very little systematic monitoring and evaluation of R&D undertaken by public research labs and universities. More effort needs to be undertaken to clarify the objectives of research in public institutions. Most important, the results of monitoring and evaluation should be used to channel resources to programs that produce results and to terminate programs that do not.

Strengthen the commercialization of knowledge

- *Improve the National Institute for Intellectual Property (INPI) by reducing the backlog of patent and trademark applications and by providing more assistance to Brazilian innovators.* Although patenting activity has increased in recent years, INPI needs to eliminate the growing backlog of applications. In addition, the enforcement of intellectual property rights needs to be strengthened so that firms will expend the effort needed to develop new knowledge.

- *Implement the Innovation Law.* As can be seen through its strong production of scientific and technical papers, Brazil does better in producing basic knowledge than in applying knowledge. Even knowledge that is patented is often not exploited for productive purposes. Part of the problem is that most of the research is done in government labs and universities, which have few incentives to exploit knowledge through commerce. The Innovation Law passed in 2005 goes some way toward allowing research institutes to commercialize knowledge that has been developed with public resources. The regulations to implement the Innovation Law have not yet been passed, so the law's full effects have yet to be felt. Some requirements, such as demanding competitive bidding for the sale of licenses, may be too onerous. The law needs to be modified to give greater stimulus to the commercialization of knowledge.

- *Support technology transfer offices in public universities and R&D institutes, as well as a patent management corporation.* This would send a powerful signal to the productive sector about the importance of adapting research for applied purposes. Exchanging experiences through an association of technology commercialization centers could help to generate economies of scale.

- *Promote greater mobility between public research personnel and the productive sector.* At present, the bulk of scientific and technical talent in Brazil resides in the university sector. Scientists and engineers are unlikely to move between the academic sector and industry, or even between the academic sector and government research labs. International experience shows that mobility serves to cross-fertilize research settings and to increase productivity. To stimulate such interaction, special programs should be funded to help subsidize the cost of personnel exchanges.

- *Expand technology parks and incubators.* The most successful of Brazil's relatively rare technology parks and business incubators are in the states of São Paulo and Rio de Janeiro. More should be set up. It will be important to avoid the common mistake of focusing too narrowly on real estate and equipment at the expense of the "soft elements" for such centers. The needs are for training in entrepreneurship for scientists and engineers (for example, pairing them with business experts), assistance to develop business and marketing plans, access to early-stage innovation finance and venture capital, assistance in protecting intellectual property, and general help in setting up and "growing" businesses.

Improve financial support for early stages of technology development

- *Improve finance and procedures for evaluating projects, and speed up approvals.* Brazil has a long history of financing early-stage technology development through institutions such as FINEP (*Financiadora de Estudos e Projetos*). However, the procedures for evaluating and approving projects need to be made more efficient and faster.

- *Improve monitoring and evaluation of ongoing projects.* More effort also has to be put into monitoring and evaluating ongoing projects to spot some of the pitfalls to be avoided, identify timely assistance to be provided, and improve future project selection and funding.

Deepen early-stage venture capital

The early-stage venture capital industry in Brazil is small and incipient. Several actions need to be taken:

- *Strengthen the supply of technology commercialization projects.* International experience shows that the initial constraint to developing a venture capital industry is the limited number of good projects. Creating a critical mass of viable "deals" requires entrepreneurial training of scientists and engineers as well as a commercially oriented approach to research.

- *Strengthen techno-entrepreneurship training in universities.* It is important to provide techno-entrepreneurial training within engineering and business schools. This type of training is poorly developed in Brazil. It needs to be strengthened.

- *Introduce regulations that facilitate the growth of venture capital.* International experience has taught that the attractiveness of venture capital investments often depends on how gains and losses will be taxed. Brazil has made recent progress in this area; however, more can be done to increase the attractiveness of providing risk capital for new start-ups.

Acquiring Foreign Knowledge

Both the country-level analysis in chapter 4 and the firm-level analysis in chapter 5 showed that Brazil is taking less advantage of global knowledge than its main economic competitors. At the macro level, this is revealed most clearly in the low share of trade in GDP and, in particular, the low level of capital goods imports. It also is seen in the relatively lower payments for technology licensing as a share of GDP.

At the micro level, the relative underutilization of foreign knowledge is reflected by Brazilian firms' low level of technology licensing. Our micro-level data confirms that new machinery and equipment are generally the firms' primary source for new technology. Low import of capital goods and low use of foreign technology reinforce each other.

The firm-level microanalysis confirmed that exporting firms are more likely to invest in R&D and to innovate than nonexporting firms. Here, causality tends to run in both directions. Innovative firms that do R&D are more likely to be competitive, and hence more likely to export. But involvement in export means that firms have to be more innovative because keeping up with foreign competitors means keeping up with their advances. In addition, the effect of foreign knowledge seems to be indirect. Domestic firms with relatively more foreign participation (either through ownership shares or product purchases) tend to benefit from a positive externality— they are more likely to become involved in R&D, they innovate, and they are focused more consistently on quality improvement. These features have important implications that help to shape the following recommendations about acquiring foreign knowledge:

Use trade to improve access to foreign knowledge

- *Expand openness to trade and to FDI flows.* The first and arguably most important recommendation is for Brazil to continue opening its trade regime to foreign competition. Despite the reforms of the early 1990s, Brazil ranks among the world's most protectionist countries in tariff and nontariff barriers; and both are particularly high for capital goods, compounding Brazil's already limited access to embodied technology. Furthermore, although Brazil has received much FDI, most of it has been oriented toward the protected domestic market rather than (like China) toward building an export platform to the world. Because of the less-demanding domestic competitive environment, foreign firms also may not be required to bring their most advanced technologies to Brazil. In China, they do—precisely because foreign firms' competition for the Chinese domestic market is more domestic. Thus, by liberalizing its trade regime, Brazil will receive triple advantage—first, lower cost for technology embodied in capital goods and components; second, more foreign products and services available for copying, reverse engineering, and technology upgrading; and third, FDI serving as an entry vehicle for advanced technology with the possibility of positive spillovers.

- *Continue to ease technology transfers.* INPI needs to reduce remaining restrictions on technology licensing; and Brazilian firms need flexibility to structure the best deals that they can get. Smaller firms can be assisted in contract negotiations rather than having to navigate rules and regulations entirely by themselves.

Support explicit acquisition of knowledge abroad
In addition to further opening its trade regime to bring in more foreign knowledge, Brazil should launch programs that proactively seek out and acquire foreign knowledge, as its Asian competitors do. It should take the following actions:

- *Set up a program to foster international research collaborations for the private and public sectors.* At present, there is scant research cooperation between foreign and domestic institutions (either public or private). Government programs should explicitly encourage such cooperation. For example, the Bird program in Israel and India is a government-supported fund to foster international research cooperation between firms to develop and commercialize new technologies.

- *Purchase foreign companies.* Japanese, Korean, Taiwanese, and more recently, Chinese and Indian firms are aggressively accessing knowledge by buying up foreign high-technology firms. The Brazilian government and private sector should emulate this strategy.

- *Purchase foreign R&D labs abroad.* Brazil's developing-country competitors also are buying foreign research institutes. Even when they do not purchase research institutes (or cannot purchase universities), they actively acquire technology through contracts and joint research endeavors. Brazil should do the same.

- *Send more Brazilian students to study abroad.* Brazil sends relatively few students for education abroad—far different from China, India, Korea, Taiwan, Malaysia, and many other countries. Foreign training provides people at the beginning of their careers with direct access to the cutting-edge of technical knowledge; and when advanced training is combined with hands-on work experience in firms, research institutes, and universities, it is an excellent way to "nationalize" commercially relevant skills. Brazil should expand programs to send students, particularly postgraduates, for education and training abroad.

- *Promote interactions and faculty exchanges with foreign universities.* Brazil's East Asian competitors continuously arrange university and faculty exchanges with the best foreign universities. The three main Chinese universities—Beijing University, Tsinghai University, and Fusan University—sponsor several hundred programs each with foreign universities. In addition, they constantly arrange programs with advanced foreign firms for training and pilot programs that test new technologies. By contrast, Brazilian universities sponsor many fewer formal programs and faculty exchanges. A change may require greater emphasis on the use of English among students, professors, and researchers.

- *Tap talent from the "Brazilian Diaspora."* Besides sending many more of their students abroad, Brazil's competitors also have developed programs to bring their trained students home. Programs of this sort include recruitment missions, generous repatriation incentives, and even special high-technology industrial parks that are aimed specifically at capitalizing on knowledge gained abroad.

Leveraging Existing Technologies

In addition to ramping up its capacity to create and acquire new technologies, Brazil needs to more productively use existing technologies. This may well be the most fruitful means to accelerate Brazil's future growth. The Korean and Chinese experiences demonstrate the importance of adopting, adapting, and effectively using existing knowledge, especially when it can be leveraged through a workforce with abundant basic skills. Companies that understand the importance of new technologies—and that have workers who can learn quickly and put these technologies to use—are in a strong position to expand their capital (in the sense of TFP) through dramatic expansion in productivity.

Although healthy mature economies typically do all three—create new technology, acquire technology from elsewhere, and make better use of the technology they already have—the "low-hanging fruit" for Brazil is in better use of existing technology. As shown with an econometric model in chapter 5, companies within the same sector could increase their output by orders of magnitude if they were to emulate best local practice. In other words, Brazilian firms not only are not producing optimally at international standards, they are not producing optimally at national standards.

More effort has to be put into upgrading technology across the economy as a whole, but in particular among the small and medium-size enterprises that compose the majority of Brazilian firms. Important initial steps have already been taken with the recent tax reform measure known as "*Lei do Bem*" and the General Law regulating micro and small enterprises, *Lei Geral da Micro e Pequena Empresa*. However, further interventions are needed as well:

Promote diffusion of technical information

- *Improve technology information services.* With the proliferation of Internet databases and advisory services, access to technical information is easier than ever. Larger firms are typically proficient at accessing and using information, but small and medium-size firms will need assistance. Strengthening their access means enhanced efforts to package information together with well-targeted advisory services. Efforts along these lines have been launched through SEBRAE (Brazilian Service for Assistance to Small Business, *Serviço Brasileiro de Apoio às Micro e Pequenas Empresas*) and industry associations; however, much more needs to be done.

- *Strengthen technology extension in agriculture, industry, and the service sectors.* It is often necessary to take information and make it meaningful by developing concrete demonstration projects—in other words, showing what has to be done, how, and to what benefit. Concrete demonstration projects are critical to attracting early adopters whose success can lead to diffusion and replication throughout the economy. Brazil has had success in agricultural extension through Embrapa, as well as state-level research and extension services. However, it has done less well with parallel efforts in industry and the services sector. There is great potential for action through government support in these areas.

Improve the diffusion and absorption of metrology, standards, testing, and quality-control (MSTQ) services

Quality is as essential as price for competitiveness in today's global economy. This means a good physical MSTQ infrastructure, as well as a culture of quality.

- *Create a world-class, demand-responsive MSTQ infrastructure.* Brazil currently has a large public infrastructure for basic measurement. Yet, Brazil's domestic norms and standards need to be assessed against international norms and standards, especially where international accreditation eventually may be at issue. This complex subject matter requires considerable technical analysis that is beyond the scope of this report. However, a more detailed study should be undertaken to identify links that may be missing within the system. Eliminating deficiencies and seeking international accreditation will be important to correcting and enhancing Brazil's international competitiveness.

- *Promote quality control in firms, encouraging them to set up labs and to seek quality certification.* Beyond the physical and regulatory infrastructure, it is necessary to create a culture of quality in the economy. The micro-evidence presented in chapter 5 of this report showed clear positive associations among quality certification, R&D innovation, and exports. This implies the need to disseminate information about the importance of quality for innovation and competitiveness. In addition, firms require help in setting up the physical infrastructure and implementing the procedures for internationally recognized quality certification. One of the special technology funds focuses on university-based research infrastructure. This support could be expanded for testing and quality control in firms. Sources of support should be explored.

Strengthen finance and training for technology absorption by small and medium enterprises (SMEs)

As discussed, a strong link has been found between firm size and innovation inputs (such as R&D efforts, skilled workers, use of computers, and purchase of technology), innovation outputs (such as product and process innovations), and outcomes (such as productivity and growth). In addition, the very large dispersion of productivity within virtually any industrial sector in Brazil confirms the very wide range of technological capability in the country. Because larger firms are likely to already be efficient, public actions are primarily needed to support the needs of SMEs, helping them to make efficient use of both acquired and existing technologies.

- *Develop support mechanisms for industrial clusters, focusing on design as well as on technological and marketing capabilities.* Work with industrial clusters is a key mechanism for improving the productivity of sectors. There is great value in sharing knowledge about the key constraints and opportunities faced by firms in specific clusters. Collective action to share information and experience in specific regional clusters is useful in identifying firms' shared constraints, threats, and opportunities. Many of these go beyond individual companies; potential solutions may need to be addressed by the group. For example, firms may band together to improve designs or to receive process consultancy assistance. Similarly, they may band together to get technical input from specialized suppliers. The group may need assistance such as a common processing facility, a quality-testing center, a market study, or a distribution system whose scale exceeds the capacity of a single firm.

- *Provide greater support for cluster diagnosis and identification of ways to improve performance.* This can often be accomplished as individual cluster members become accustomed to working as a club. Some initial government incentive—for example, subsidizing the cost of the initial diagnosis or initial purchase of expertise—may be necessary to catalyze group information sharing and joint action. In addition, changes in state or local actions—such as specialized training institutions, better transportation, communications infrastructure, or finance—may be necessary to the solution; so state involvement may be required.

- *Strengthen finance for technology absorption by SMEs.* The availability of finance is typically the most important constraint for SMEs. This is particularly important in Brazil where the cost of capital is very high. Thus, it is important to focus not only on how to improve technical information, but also on how to invest in better equipment and inputs. With costly financing, it obviously makes sense to first focus on improvements in product, process, and quality (which require less new investment). However, other kinds of improvements—for example, buying better equipment or upgrading worker skills—may still make sense, the high cost of capital notwithstanding.

Improve Basic Education and Skills

Brazilian firms frequently are required to train their employees in basic math and reading, crowding out technical job training that could more directly increase productivity. The country's educational system, not employers, should be responsible for basic education. The World Bank has produced many studies on policy options to address shortcomings in the basic and tertiary education systems.[4] Here, we group recommendations in four main areas—governance, quality, access to secondary education, and school-to-work transitions.

Governance

Introduce a performance-based culture
Clear division of responsibility is particularly important in a federal political system. While distribution of responsibilities is defined in Brazil's education sector, overlap, role conflicts, and inefficiencies are common. The Ministry of Education is supposed to avoid intervention as a primary deliverer of educational services. That is the job of the states and municipalities. A ministry's role is to set performance goals and provide incentives to help the states and municipalities meet them. As reiterated in the Lula administration's recently launched Education Development Plan (*Plano de Desenvolvimento da Educação*, PDE), financial incentives are supposed to transition from rewarding higher enrollments to rewarding stronger performance. States and municipalities are expected to operationalize performance goals—in particular, by holding principals and their staffs accountable for achievement. The necessary school autonomy can be strengthened in several ways.

Strengthen assessment systems that measure progress and that value monitoring and evaluation
Over the past 15 years, Brazil has gained considerable experience with testing in basic education (SAEB), secondary education (ENEM), adult education (ENCEJA), and tertiary education (ENADE). At each educational level, testing tools have been developed to provide snapshots and analyze trends in student learning. These federally implemented tests are sample-based, so some states have developed their own universal testing

systems. In 2005, the Ministry of Education administered the Prova Brasil, a US$25 million learning assessment of 3.3 million basic education students in over 42,000 schools. The incipient "culture of evaluation" must be preserved and deepened, at the same time avoiding redundant testing at the multiple levels of governance.

Quality

Low educational quality is associated with high repetition and dropout rates. It is easy to see how underfunding and poor use of resources reinforce a vicious cycle of low quality and high repetition. The following paragraphs present some of the policy options to help break this cycle.

Establish minimum operational standards for schools and municipal secretariats

Brazil is well aware of what municipal secretariats and classrooms need to effectively manage schools and improve student learning. Yet many schools—especially those in rural areas of the poorest regions—still lack suitable physical classrooms, basic furniture, and teaching materials. The past decade has seen enormous progress in institutionalizing minimum operational standards, yet much of this task still remains to be done.

Retrain teachers and reward performance

Most Brazilian teachers receive their degrees from small private universities of highly uneven quality. In general, they arrive in the classroom trained in philosophical aspects of pedagogy rather than with practical strategies on how to teach. They typically come with very few tools and even less experience in managing classrooms. Incentives and opportunities for retraining teachers and rewarding performance are badly needed across the system.

- *Reward good teaching with tangible incentives and punish the absenteeism that is particularly flagrant in many rural schools.* High-performing teachers (and teachers who improve) should be publicly recognized by the Ministry of Education and the corresponding state or municipal secretariat. Rewards can be allocated individually or collectively to a school. At the same time, studies have confirmed the pervasiveness and high cost of continuous teacher absence in many municipalities. Patterns of abuse need to be detected and punished.

- *Revise the professional advancement structure.* Promotions and career advancement should be linked to performance rather than to seniority or attendance at training courses. Recent research by Universidade de São Paulo professor Naercio Menezes Filho confirms that the current training courses generally are not focused on student learning, and attendance is not a good predictor of improved classroom performance.

Recruit the best candidates into teaching

Offer grants to prospective teachers that will support them through their tertiary studies. Make the teaching profession attractive to more candidates and thus more competitive and selective. Set entry salaries higher and decrease the salary gradient over a professional career. If the right candidates are attracted to the classroom, those with a true vocation will remain through retirement.

Select competent, certified school principals

Virtually all research confirms the critical importance of school principals in shaping the success of schools. Today, some researchers estimate that more than 60 percent of Brazilian principals obtain their jobs based on political criteria. They need to be selected based on competence instead, preferably following a certification process that ensures their pedagogical competence and administrative skills to manage a school.

Build upon the existing school councils, strengthen the relationships between schools and communities

Brazil has a long history of community participation in schools. It is important to energize schools by building upon that foundation—bringing in speakers from the community, organizing school events open to the community, and creating internship opportunities for graduating students in local industries and firms.

Invest in early childhood education

International research shows that entering first graders from poor families generally know about 400 words, compared with the 4,000 words or so known by first graders from the top economic quintile. Even the highest-performing schools will struggle to overcome an initial disadvantage of this sort. A strong preprimary experience strengthens students' readiness. Preprimary education in Brazil is generally the responsibility of municipalities. The approval of FUNDEB, which includes funding for preprimary education, offers a promising opportunity that requires strong oversight and support from the federal and state governments.

Access to Secondary Education

Europe and the United States provide the two major models for secondary education. In Europe, schools offer distinct educational modalities, each serving a particular student profile—some mostly academic, others technical/professional, others purely vocational. The American model offers just one type of school, with students usually able to select vocational courses to complement a core academic track. Currently, Brazil's schools more closely resemble the American model, though with even fewer vocational courses to choose from. Basically, all students are placed on an academic track, with 55 percent of enrollees attending evening shifts. Because primary-education quality is frequently so poor, many students acquire their basic skills functionally while attending at the secondary level. Brazil's model may have to adjust over time, but it is probably appropriate to the reality and needs of its students in the meantime.

Improve secondary schools by improving primary schools

The expansion of secondary education depends not just on additional financing (for which FUNDEB will be critical) but on stabilizing the flow of students arriving from primary schools. Recent budgetary increases for secondary education mean that expansion of coverage is likely to accelerate. That, however, is not enough: Students must complete the cycle. Overwhelmingly, those who drop out are also those who repeated early years (additionally complicating schooling through age/grade distortion). For this reason, the key to success at the secondary level is to improve quality and decrease repetition in the lower grades. Efficiency gains at the primary level can also help to finance secondary expansion. With the cost of repetition estimated at US$600 million, it is imaginable that significant savings could be transferred. The shift in resource allocation would likely occur at the state level, which would need to be monitored in order to avoid harmful reductions in per pupil expenditures at the primary level.

Use conditional cash transfers (CCTs) and savings accounts to help attract secondary students to school and to help keep them there until graduation

The expansion of *Bolsa Família* benefits to cover secondary-school attendance is under discussion; and new ideas are on the table, such as student savings accounts that could be accessed upon secondary-school completion. These are good ideas; however, demand-side interventions should not crowd out funds from school budgets that are otherwise urgently needed to achieve minimum operational standards for the schools that these students will attend.

Facilitation of the School-to-Work Transition

The school-to-work transition takes place for many students at the end of the secondary cycle, either in its regular format or in the adult education format (*Educação de Jovens e Adultos*, EJA). Very high youth unemployment suggests a need to strengthen the transition for students entering the workforce. The following paragraphs highlight several policy options to facilitate this transition.

Within secondary education, track students more realistically by age to better target school-to-work interventions for those who will face the job market soonest

By placing older students in the evening shift and younger students in the day shift, classes would be more homogenous, and age-appropriate curricular differentiation could be introduced. The older evening students could be placed on an accelerated basic skills curriculum similar to EJA, in which all students also receive instruction in workplace skills such as communications, computer use, and negotiation. For the younger day students, "skills for work" training can complement the academically oriented track that is otherwise geared toward preparation for the tertiary system.

Strong linkages should be built among secondary schools serving older students and employers, technical and vocational service providers, and the S-system

Technical and vocational training should be left to postsecondary education, where it should be provided primarily through short, flexible, narrowly focused courses. However, recruitment for these courses should begin early for students in the evening secondary schools.

Encourage validation exams as a means to obtaining secondary-school diplomas for older students who have learned and acquired experience through alternative methods such as employment

A validation exam already exists in Brazil; however, it is not widely used, and its application is time-bounded. The validation exam should be readily available, preferably online. It should be geared toward adults (persons older than 20) who wish to obtain a secondary-education degree after demonstrating that they have successfully mastered the competencies and knowledge that otherwise might have been learned in schools.

Expand Tertiary Education and Advanced Skills Training

The Brazilian government's ambitious plan to increase tertiary education coverage, achieve greater equity, enhance quality, and improve relevance cannot be achieved narrowly through the traditional approach of publicly funding new public universities. The next paragraphs present the policy options for improving tertiary education. The discussion is divided into policy options on governance and financing, quality and relevance, and the need to develop world-class universities.

Governance and Financing of Tertiary Education

Promote greater autonomy for institutions while simultaneously putting adequate accountability mechanisms in place

Greater autonomy and accountability will allow public universities to strengthen their performance and become more innovative. The government can help to achieve this through shared planning and setting of quantitative and qualitative goals. The Ministry of Education and the productive sectors need to jointly develop a rigorous system of results-oriented evaluation. Indicators should be clear and measurable, laying out specific institutional, academic, and financial outcomes to which all actors can be held accountable.

Make rules on resource utilization more flexible

To promote greater efficiency in the use of public resources, the government should consider a combination of complementary mechanisms for allocating funding to tertiary institutions based upon measurable performance.

Ensure adequate coverage and long-term sustainability of support, especially for low-income students

The government of Brazil needs to increase funding for low-income students while ensuring high levels of repayment. The government may wish to explore the feasibility of creating an income-contingent student loan system that, in principle, would be more efficient and equitable than the present mortgage-type scheme. The government might also consider international loans to finance an educational credit program. In this case, the funds would be channeled through an association of private schools—as was done in Mexico, for example, through a World Bank loan. In any event, mechanisms to finance students should be defined with criteria and priorities that build upon external evaluation results.

A labor-market observatory needs to be established to monitor what happens to tertiary graduates

Findings on careers and pathways should be widely disseminated. This is critical not just for informing job-seeking students, but for helping decision makers keep tertiary education and labor market policies in optimal sync.

Quality and Relevance of Tertiary Education

Focus on quality

Institutions need to raise the qualifications of their academic staff, improve pedagogical practices, integrate research into the undergraduate curriculum, upgrade their infrastructure, and provide stimulating learning environments. Close linkages must be forged with the productive sectors, especially for professional tracks and programs related to science and technology.

Focus on education first, not on research

Even in those countries with the highest degree of scientific production, nearly all universities insist on quality of education first, not research. Relatively few institutions have the vocation or resources to conduct research in every department (in the United States, for example, only 3 percent to 5 percent of universities are classified as "research universities"). In Brazil, teaching centers—whether or not they are legally defined as universities—could and should support research centers. Research is not their primary mission; yet learning the scientific method—surely a cornerstone of what is meant when someone is called "well educated"—requires that all students conduct and apply research to some extent. Universities differ from pure R&D labs in that their objectives, at least for beginners, are primarily didactic. Less directly, this process also leads to scientific and specialized publishing, and to the capacity for productive innovation at the national level.

Place more emphasis on educating locally responsible global citizens

Tertiary education institutions in Brazil need to view their mission as preparation of globally minded, locally responsible, internationally competitive citizens. Brazil needs to improve foreign language training for both academic staff and graduates. The country would benefit from a two-way street of exchanges: facilitating international mobility for Brazilian students, professors, and researchers while also welcoming foreign professors and students to study and collaborate in Brazil. Resources should be made available to support these initiatives.

Encourage more students to enter science and engineering

A major push will be needed to train more and better scientists and engineers. At the same time, attractive nonuniversity alternatives should be developed to train middle-level professionals and technicians. Because a large percentage of lower-income students are enrolled in these courses, scholarships, education credits, and ProUni (Programa Universidade para Todos) should be encouraged for this field of study. In addition, the Ministry of Education should control the quality of courses through periodic randomized technical visits. This would complement targeted visits to courses for which there are no clear indicators of program quality.

Build strong links between top research universities and the productive sectors

A major push is also needed to promote the commercialization of knowledge and innovation already being developed by top research universities. Exchange programs between universities and the productive sectors should be supported, and links between universities and business incubators should be strengthened. Brazil should also promote the development of proof-of-concept centers, a new type of in-house business incubator offering seed funding and other services specifically targeted to university researchers. The Ewing Marion Kauffman Foundation recently published an insightful study of two U.S. proof-of-concept centers that emerged just a few years ago to accelerate the transfer of academic innovations into commercial applications in the United States (see www.kauffman.org).

World-Class Universities

The government of Brazil should decide how many world-class universities the country needs and can afford

What are the criteria for selecting and funding world-class universities in Brazil—and at what opportunity cost to the rest of the education system? If decisions are made to compete in this rarefied arena, explicit policies and substantial investment should be made to build upon the foundation provided by current centers of excellence.

Increase funding for the top research leaders

The budgets of the 10 leading research universities should reflect their productivity. In parallel, the best graduate studies programs—in both public and private universities—also should have their budgets increased. As a fraction of overall education budgets, incremental expenditures of this sort would be virtually negligible. Their signaling and productive benefits would be incalculable.

CHAPTER 8

From Analysis to Action

Innovation and economic growth are broad topics, so this report has ranged across a broad spectrum of issues—from the overarching economic and institutional regime (macroeconomic parameters, government regulation, trade and competition policy, security, and the rule of law) to specific areas (public and private R&D; foreign investment and technology transfer; technical information; metrology, standards, and quality control; education and skills; and innovation finance and venture capital). Based on this analysis, we have suggested a set of actions (chapter 7) to help Brazil become a more aggressive and successful player in the global economy. This chapter looks at the many entities of government, the private sector, and civil society that will have to implement these recommendations if ideas are to be translated into action and then into reality (table 8.1).

Who Needs to Be Involved?

Not all of the recommendations in chapter 7 (summarized in table 8.1) are of equal weight and priority; and for technical and political reasons, some will be more difficult to implement than others. Moreover, collaborative action among actors will be needed, so the key agencies outlined in table 8.1 are meant to indicate those that could lead in coordinating actions rather than those solely responsible for carrying out actions. Some recommendations would require new laws through Congress; some imply policy actions embodying significant changes in regulations; and others could be done with the mere stroke of a pen (and a great deal of political will). Some could be carried out with existing resources; others would require significant mobilization of public and private funds. Some actions are stand-alone, while others must be coordinated and sequenced with related steps. Some will require years of sustained efforts; others could be done rapidly. But overall, this report signals that a coordinated and sustained effort by the government of Brazil is urgent.

Table 8.1. Who Needs to Do What

	Recommendations in which they need to be actively involved
Federal government as a whole	• Improve governance and decrease red tape.
Ministry of Finance	• Stay the course in continuing to improve the basic macroeconomic environment.
	• Facilitate firm-level investment by lowering the cost of capital.
	• Increase private R&D by (a) expanding the mix of public finance instruments that foster private R&D to include more risk sharing, matching grants, equity instruments, and loans and (b) improving the cost-effectiveness of fiscal incentives for R&D.
	• Improve public R&D by increasing public resources for it and by improving the monitoring and evaluation of public research, using the results to reallocate funds by performance.
	• Deepen early stage venture capital by introducing regulations that facilitate the growth of venture capital.
Central Bank	• Stay the course in continuing to improve the basic macroeconomic environment.
	• Facilitate firm-level investment by lowering the cost of capital.
	• Improve the efficiency and intermediation of the banking system.
Ministry of Trade, Industry, and Commerce	• Facilitate firm-level investment by continuing to open the economy to foreign competition.
	• Strengthen the commercialization of knowledge by (a) improving the National Institute for Intellectual Property (INPI) by reducing the backlog of patent and trademark applications and providing more assistance to Brazilian innovators; (b) implementing the Innovation Law; (c) supporting technology transfer offices in public universities and R&D institutes, as well as a patent management corporation; (d) promoting greater mobility between public research personnel and the productive sector; and (e) expanding technology parks and incubators.
	• Use trade to improve access to foreign knowledge by expanding openness to trade and to FDI flows and continuing to ease technology transfers.
	• Improve the diffusion and absorption of MSTQ services by creating a world-class, demand-responsive MSTQ infrastructure and by promoting quality control in firms, encouraging them to set up labs and to seek quality certification.
	• Strengthen finance and training for technology absorption by SMEs by (a) developing support mechanisms for industrial clusters, focusing on design as well as on technological and marketing capabilities; (b) providing greater support for cluster diagnosis and identification of ways to improve performance; and (c) strengthening finance for SME technology absorption.

(continued)

	Recommendations in which they need to be actively involved
Ministry of Education	• Support explicit acquisition of knowledge abroad by (a) sending more Brazilian students to study abroad, (b) promoting interactions and faculty exchanges with foreign universities, and (c) tapping talent from the "Brazilian Diaspora."
	• Improve governance of the basic education system by introducing a performance-based culture; expand the use of tests that evaluate student learning.
	• Improve the quality of basic education by (a) encouraging the introduction of minimum operational standards in all schools, (b) strengthening the teaching force, (c) upgrading the skills of school principals, (d) increasing investment in early childhood education, (e) building upon existing school councils to increase collaboration among schools and their surrounding communities, and (f) discouraging repetition.
	• Expand access to secondary schooling by improving the student flow in primary education, and utilizing conditional cash transfers in secondary school to discourage dropouts
	• Enhance the school-to-work transition by (a) tracking students more realistically by age within secondary education to better target school-to-work interventions for those who will face the job market soonest; (b) building strong linkages among secondary schools serving older students and employers, technical and vocational service providers, and the S-system; and (c) encouraging validation exams as a means of obtaining secondary-school diplomas for older students who have learned and acquired experience through alternative methods such as employment.
	• Promote greater autonomy for institutions while simultaneously putting adequate accountability mechanisms in place by (a) making rules on resource utilization more flexible; (b) ensuring adequate coverage and long-term sustainability of support, especially for low-income students; and (c) establishing a labor market observatory to monitor what happens to tertiary graduates.
	• Increase the quality and relevance of tertiary education by focusing on education first, not research.
	• Place more emphasis on educating locally responsible global citizens, on encouraging more students to enter science and engineering, and on building strong links between top research universities and the productive sectors.
	• Develop world-class universities. The Brazilian government should decide how many world-class universities the country needs and can afford.
Ministry of Science and Technology and FINEP	• Increase private R&D by (a) expanding the mix of public finance instruments that foster private R&D to include more

(continued)

Table 8.1. (*continued*)

	Recommendations in which they need to be actively involved
	risk sharing, matching grants, equity instruments, and loans; (b) improving the monitoring of how well different instruments and mechanisms work; (c) making R&D fiscal incentives more cost effective; (d) improving operation of the sectoral science and technology funds to provide greater flexibility across sectors and to increase interaction among academia, research institutes, and private firms.
	• Improve public R&D by (a) increasing public R&D resources; (b) strengthening public R&D in key strategic areas, such as natural resources, renewable energy, biotechnology, and nano-technology; (c) increasing support for R&D in universities; and (d) improving monitoring and evaluation of public research, using the results to reallocate funds by performance.
	• Strengthen the commercialization of knowledge by (a) implementing the Innovation Law; (b) supporting technology transfer offices in public universities and R&D institutes, as well as a patent management corporation; (c) promoting greater mobility between public research personnel and the productive sector; and (d) expanding technology parks and incubators.
	• Improve financial support for early stage technology development by (a) improving finance and procedures for evaluating projects and speeding up approvals and (b) improving monitoring and evaluation of ongoing projects.
	• Support explicit acquisition of knowledge abroad by (a) setting up a program to foster international research collaborations for the private and public sectors, (b) purchasing foreign R&D labs abroad, and (c) tapping talent from the "Brazilian Diaspora."
	• Improve the diffusion and absorption of MSTQ services by (a) creating a world-class, demand-responsive MSTQ infrastructure; and (b) promoting quality control in firms, encouraging them to set up labs and to seek quality certification.
	• Strengthen finance and training for technology absorption by SMEs by (a) developing support mechanisms for industrial clusters, focusing on design as well as technological and marketing capabilities; (b) providing greater support for cluster diagnosis and identification of ways to improve performance; and (c) strengthening finance for SME technology absorption.
FAPESP and other R&D state-level agencies	• Improve public R&D by (a) strengthening public R&D in key strategic areas, such as natural resources, renewable energy, biotechnology, and nanotechnology, and (b) improving monitoring and evaluation of public research, using the results to reallocate funds by performance.
	• Strengthen the commercialization of knowledge by (a) supporting technology transfer offices in public universities and

(*continued*)

	Recommendations in which they need to be actively involved
	R&D institutes, as well as a patent management corporation; (b) promoting greater mobility between public research personnel and the productive sector; and (c) expanding technology parks and incubators.
	• Improve financial support for early stage technology development by (a) improving finance and procedures for evaluating projects and speeding up approvals and (b) improving monitoring and evaluation of ongoing projects.
	• Strengthen finance and training for technology absorption by SMEs by (a) developing support mechanisms for industrial clusters, focusing on design as well as technological and marketing capabilities; (b) providing greater support for cluster diagnosis and identification of ways to improve performance; and (c) strengthening finance for SME technology absorption.
BNDES	• Increase private R&D by expanding the mix of applicable public financial instruments to include more risk sharing, matching grants, equity instruments, and loans.
	• Improve public R&D by (a) increasing public R&D resources; (b) strengthening public R&D in key strategic areas, such as natural resources, renewable energy, biotechnology, and nanotechnology; and (c) improving monitoring and evaluation of public research, using the results to reallocate funds by performance.
	• Strengthen the commercialization of knowledge by expanding technology parks and incubators.
	• Improve financial support for early stage technology development by (a) improving finance and procedures for evaluating projects and speeding up approvals and (b) improving monitoring and evaluation of ongoing projects.
CAPES	• Support explicit acquisition of knowledge abroad by (a) sending more Brazilian students to study abroad, (b) promoting interactions and faculty exchanges with foreign universities, and (c) tapping talent from the "Brazilian Diaspora."
Ministry of Justice	• Modernize intellectual property laws and strengthen their enforcement.
States	• Facilitate firm-level investment by addressing infrastructural deficiencies that increase the costs of production and of doing business.
	• Increase the use of public-private partnerships to amplify leverage investments in infrastructure.
	• Improve governance and decrease red tape.
	• Increase private R&D by (a) expanding the mix of applicable public finance instruments to include more risk sharing, matching grants, equity instruments, and loans and (b) improving the monitoring of how well the different instruments and mechanisms work.

(continued)

Table 8.1. (*continued*)

	Recommendations in which they need to be actively involved
	• Improve public R&D by (a) increasing public R&D resources; (b) strengthening public R&D in key strategic areas, such as natural resources, renewable energy, biotechnology, and nanotechnology; (c) increasing support for R&D in universities; and (d) improving monitoring and evaluation of public research, using results to reallocate funds by performance.
	• Strengthen the commercialization of knowledge by (a) promoting greater mobility between public research personnel and the productive sector and (b) expanding technology parks and incubators.
	• Strengthen finance and training for technology absorption by SMEs by (a) developing support mechanisms for industrial clusters, focusing on design as well as technological and marketing capabilities; (b) providing greater support for cluster diagnosis and identification of ways to improve performance; and (c) strengthening finance for SME technology absorption.
	• Improve governance of the basic education system by (a) introducing a performance-based culture and (b) expanding use of tests that evaluate student learning.
	• Improve the quality of basic education by (a) introducing minimum operational standards in all schools, (b) strengthening the teacher force, (c) upgrading the skills of school principals, (d) building upon existing school councils to increase collaboration among schools and their surrounding communities, and (e) discouraging repetition.
	• Expand access to secondary schooling by (a) improving the student flow in primary education and (b) utilizing conditional cash transfers in secondary school to discourage dropouts.
	• Enhance the school-to-work transition by (a) tracking students within secondary education more realistically by age to better target school-to-work interventions for those who will face the job market soonest; (b) building strong linkages among secondary schools serving older students and employers, technical and vocational service providers, and the S-system; (c) encouraging validation exams as a means of obtaining secondary-school diplomas for older students who have learned and acquired experience through alternative methods such as employment.
Municipalities	• Facilitate firm-level investment by addressing infrastructural deficiencies that increase the costs of production and of doing business.
	• Increase the use of public-private partnerships to amplify leveraged investments in infrastructure.
	• Improve governance and decrease red tape.

(*continued*)

	Recommendations in which they need to be actively involved
	• Improve governance of the basic education system by introducing a performance-based culture and by expanding use of tests that evaluate student learning.
	• Improve the quality of basic education by (a) introducing minimum operational standards in all schools, (b) strengthening the teaching force, (c) upgrading school principals' skills, (d) raising investment in early childhood education, (e) building upon existing school councils to increase collaboration between schools and their surrounding communities, and (f) discouraging repetition.
	• Expand access to secondary schooling by improving the student flow in primary education.
Private firms	• Increase private R&D by improving interaction among public labs, universities, and the productive sector.
	• Strengthen the commercialization of knowledge by (a) promoting greater mobility between public research personnel and the productive sector and (b) expanding technology parks and incubators.
	• Deepen early stage venture capital by (a) strengthening the supply of commercial technology projects and (b) strengthening techno-entrepreneurial training in and with universities.
	• Support explicit acquisition of knowledge abroad by (a) setting up a program to foster international research collaborations for the private and public sectors, (b) purchasing foreign companies, (c) purchasing foreign R&D labs abroad, and (d) tapping talent from the "Brazilian Diaspora."
CNPq	• Support explicit acquisition of knowledge abroad by (a) setting up a program to foster international research collaborations for the private and public sectors, (b) sending more Brazilian students to study abroad, (c) promoting interactions and faculty exchanges with foreign universities, and (d) tapping talent from the "Brazilian Diaspora."
Ministries conducting R&D	• Increase private R&D by (a) expanding the mix of applicable public financial instruments to include more risk sharing, matching grants, equity instruments, and loans and (b) improving the monitoring of how well different instruments and mechanisms work.
	• Improve public R&D by (a) strengthening public R&D in key strategic areas, such as natural resources, renewable energy, biotechnology, and nanotechnology and (b) improving monitoring and evaluation of public research, using results to reallocate funds by performance.

(continued)

Table 8.1. (*continued*)

	Recommendations in which they need to be actively involved
Ministry of Energy	• Promote diffusion of technical information by (a) improving technology information services and (b) strengthening technology extension in agriculture, industry, and the service sectors. • Facilitate firm-level investment by addressing infrastructural deficiencies that increase the costs of production and of doing business. • Increase the use of public-private partnerships to amplify leveraged investments in infrastructure.
Ministry of Transportation	• Facilitate firm-level investment by addressing infrastructural deficiencies that increase the costs of production and of doing business. • Increase the use of public-private partnerships to amplify leveraged investments in infrastructure.
Ministry of Telecommunications	• Facilitate firm-level investment by addressing infrastructural deficiencies that increase the costs of production and of doing business. • Increase the use of public-private partnerships to amplify leveraged investments in infrastructure. • Promote diffusion of technical information by improving technology information services.
SEBRAE	• Strengthen finance and training for technology absorption by SMEs by (a) developing support mechanisms for industrial clusters, focusing on design as well as technological and marketing capabilities; (b) providing greater support for cluster diagnosis and identification of ways to improve performance; (c) strengthening finance for SME technology absorption.
S-system agencies	• Enhance the school-to-work transition by building strong linkages among secondary schools serving older students and employers, technical and vocational service providers, and the S-system. • Increase the quality and relevance of tertiary education by (a) emphasizing the education of locally responsible global citizens, (b) encouraging more students to enter science and engineering, and (c) building strong links with top research universities and the productive sectors.
Ministry of Labor	• Facilitate firm-level investment by challenging the rigidity of Brazil's labor markets.
Public universities and labs	• Increase private R&D by improving interaction among public labs, universities, and the productive sectors. • Strengthen the commercialization of knowledge by (a) supporting technology transfer offices in public universities and R&D institutes, as well as a patent management corporation; (b) promoting greater mobility between public research personnel and the productive sector; and (c) expanding technology parks and incubators.

(*continued*)

Recommendations in which they need to be actively involved
• Deepen early stage venture capital by (a) strengthening the supply of commercial technology projects and (b) strengthening techno-entrepreneurial training in universities.
• Support explicit acquisition of knowledge abroad by (a) setting up a program to foster international research collaborations for the private and public sectors, (b) promoting interactions and faculty exchanges with foreign universities, and (c) tapping talent from the "Brazilian Diaspora."

Source: Author.

Next Step—Raising Awareness

This report takes a first step in moving beyond analysis toward a concrete plan. Because so many institutions and actors will need to be involved, the first and most urgent action is to build awareness of the challenge. Nothing less will suffice than a fundamental change in national mindset.

Many national magazines have published article series that have helped heighten public interest in the issues discussed in this report. Among others, the magazines include *Veja, Exame,* and *Época*; and the newspapers include *Folha de São Paulo, Estado de São Paulo, O Globo, Correio Braziliense,* and *Valor Economico.* Additional public interest has been raised through films, documentaries, radio, and television. To build upon this public interest, seminars could be offered through the Brazilian Congress that involve government ministries and major civil society organizations, including labor unions and private sector associations. (For a good example of how another country took on this challenge, see box 8.1 on the Vision Korea Project.)

Following awareness building, concrete action plans need to be developed and then implemented. These need not be fully integrated plans that tackle all issues at once, as was done in Korea, but they do need to be concrete and explicit about where and with whom to start. Some actions at the federal level should rightly be top-down—in particular, measures applied at the macroeconomic level to enable growth from below. Many enabling conditions are better expedited at the state level—the process of getting a permit to start a business, provision of infrastructure services, basic and secondary education, skills training, and so forth. In other cases, bottom-up actions need to percolate from states, regions, clusters, or even organizations. The key is to get the process moving both from both the top down and from the bottom up. Successful bottom-up actions can powerfully demonstrate ideas that can be replicated and scaled up. Box 8.2 provides an illustration of this kind of process—the major policy changes made during China's very successful trade reforms.

Many examples illustrate how countries have made large gains by pursuing strategies involving both top-down and concrete bottom-up actions. Too frequently the main impetus is a major crisis, as happened in Finland, Ireland, and Korea. As the case of China demonstrates, however, it is also possible to experiment with changes even without a crisis.

Box 8.1. The Vision Korea Project—A Bottom-Up Initiative that Led to Government Action

In 1998, the Republic of Korea officially launched a national strategy to move to a knowledge-based economy in the wake of a financial crisis. The impetus came from the private sector—the *Maeil* business newspaper. In 1996, even before the crisis, the paper argued for a more coherent vision of the future of the Korean economy. The newspaper owner contracted the consulting firm Booz Allen Hamilton to undertake a study of the vulnerability of the Korean economy to a financial crisis like that in Mexico at the end of 1994. The *Maeil* convened a national conference to discuss the economic vulnerability that it found. The paper launched the Vision Korea Project as a national campaign in February of 1997 and commissioned a second consultancy study by McKinsey to underpin it. The 1997 Asian financial crisis, which also severely affected Korea, occurred when the report was still in process. This report found that Korea was caught in a "nutcracker" between low-wage competition from China that was quickly moving into higher-technology production and technology-based competition from Japan, Europe, and the United States. When the study was completed, the newspaper convened a second national conference to discuss the findings. Awareness of the need to change Korea's development strategy began to build among government, business leaders, and civil society at large.

Not satisfied with simply changing attitudes, the newspaper contracted a third consultancy firm to go beyond diagnostics and develop a concrete proposal for action. This study was done by the Monitor consultancy company and was titled "From Knowledge to Action." When this report was completed, the newspaper convened a large national conference to which the president of Korea, key ministers, and representatives from the private sector and civil society were invited. This conference was instrumental in getting the government to change strategy to become a knowledge-based economy.

Eventually, the government—the Ministry of Finance and Economy—became the main champion of the knowledge economy policy agenda. It contracted Korea's premier think tank, the Korean Development Institute, to coordinate the work of a dozen think tanks. A joint World Bank and OECD report provided a framework, outlining concrete steps for reforms in the various policy domains.

Progress was monitored closely. This was a crucial step in identifying and addressing any inertia or resistance, as for example, within the education system. Korea's knowledge strategy of April 2000 evolved into a three-year action plan for five main areas: information infrastructure, human resources, knowledge-based industry, science and technology, and elimination of the digital divide. To implement the action plan, Korea established five working groups involving 19 ministries and 17 research institutes, with the Ministry of Finance and Economy coordinating implementation. Every quarter, each ministry submits a

(continued)

Box 8.1. (*continued*)

self-monitoring report to the Ministry of Finance and Economy, which puts out an integrated report detailing progress. The midterm results and adjustments to the plan are sent to the executive director of the National Economic Advisory Council, which reports on the progress of implementation and gives an appraisal of the three-year action plan to its advisory members.

Box 8.2. Export Processing Zones and Trade Reform in China

China's trade reform started with the creation of four special export processing zones along the coast and eventually expanded to 19 zones as they proved to be successful. The growth of jobs and foreign exchange was spectacular, and it led to massive migrations of people from the rural areas to the coastal zones. The zones were expanded further. Seeing the benefits from greater insertion into the global trading system, China eventually decided to join the WTO and to significantly reduce its trade barriers while continuing to improve its broader enabling environment. The success of that strategy is seen in how China has been profiting from integration into the global system.

Although slow growth has generated concern, Brazil today is not faced with the kind of obvious crisis that mobilizes public concern and generates public outcry for remedial action. From within Brazil, it is not always fully apparent how demanding and intensely competitive the international economic environment has become—and the extent to which the country, despite its size and many accomplishments, is starting to lag farther and farther behind an increasing number of countries. In general, Brazilians have been far too slow in recognizing that the path forward will become increasingly steep unless significant reforms are undertaken—and that those reforms must begin now.

This report has attempted to diagnose the main challenges and outline a necessary course of action. Not only does this analysis need to ·be deepened, but as Korea among others has demonstrated, analysis always needs to be linked to decisive action. A high-level task force is required to bring analysts and policy makers together with business and social leaders. As emphasized above, this action planning must be undergirded by a broad strategy to build public awareness of what is at stake and to mobilize support for beginning to tackle the larger problems.

In contrast to Korea, Brazil is a heavily decentralized country; thus, many actions will have to occur at the municipal or state level. For this reason, the same diagnostic framework that was applied for Brazil as a nation can be applied to states or large municipalities. However, the enormous differences among states justifies drilling down more precisely to the state and large-municipality level because different mixes of innovation policies likely are more appropriate for different levels of development. In the next section, we discuss the key elements of this framework and how they could be applied in interested subnational entities.

Applying the Framework in Subnational Entities

The framework applied in this report assesses the strengths and weaknesses in four major areas of analysis: (a) the enabling environment for innovation, (b) the ability to create and commercialize knowledge (with a national-level and firm-level analysis), (c) the ability to acquire and absorb knowledge from abroad (with a national-level and firm-level analysis), (d) the ability to disseminate and use knowledge that exists in-country (with a national-level and firm-level analysis), and (e) the mechanisms and institutions in place to develop human capital (basic and advanced skills) for innovation.

Any application of this framework to a subnational entity will begin with a detailed analysis of the growth trends and composition in the entity, always using international comparative indicators and Brazilian state-level comparisons. This includes a profile of the comparative productive advantages of the geographical region and the implications these advantages hold for future growth and development. These comparative productive advantages are critical, since the main gains in competitiveness lie in stepping up innovation in precisely those processes and products.

The analysis must still return to the enabling environment for innovation and growth, with the clear understanding that some elements will be exogenous to the subnational entity (such as exchange and interest rates) because they are parameters set either by international markets or by federal authorities. However, the analysis also should identify elements of the enabling environment that are affected by the subnational government's policies (for example, physical infrastructure, good governance, lower crime, and reduction of red tape) and should propose options to magnify their beneficial impact on innovation and growth.

In analyzing an entity's ability to create and commercialize knowledge and technology, its R&D efforts and impact must be studied, whether financed by the public or the private sector. The analysis must study the determinants that explain why and when a firm engages in R&D, and which sectors are more likely to be involved in the kind of subnational entity being studied. Analysis should also examine the instruments and mechanisms available to facilitate interaction between firms and universities. Here, once again, there must be a clear understanding that some elements will be exogenous to the subnational entity (such as the content of national laws, especially the Innovation Law) because they are parameters set either by international markets or by federal authorities. However, the analysis should identify elements for the creation and commercialization of knowledge and technology that are affected by the subnational government's policies (for example, creation of incubators, or effectiveness of state or municipal R&D financing), and must propose options to enhance their positive impact on innovation and growth.

In analyzing the subnational entity's ability to acquire and absorb foreign knowledge and technology, the exercise must look into the export/import characteristics of the entity; the information and technology networks and connectivity; the access to ports, airports, and land transportation; and local firms' behavior regarding capital investments.

In analyzing the entity's ability to disseminate and use knowledge and technology that is already present in the system, the analysis must look at the entity's capacity in technology information services, support mechanisms to industrial clusters and production chains, and laboratories for quality certification, among other factors. The study must identify entity policies that may contribute to stronger dissemination and use of knowledge and technology.

Finally, the analysis should focus on the entity's institutional policies and performance in preparing the critical human capital needed to advance the innovation agenda. This report has established the importance of strong basic skills, especially for the absorption and diffusion of knowledge, and of advanced skills for the creation and commercialization of knowledge. The state-level study must analyze the performance of the formal education system (basic and tertiary) as well as the training offered through alternative institutions, such as the S-system and private training agencies and within firms.

Using this analytical framework, the "drill-down" work will yield specific recommendations and policy options for the subnational entity to target its efforts to strengthen and foster innovation, productivity, and economic growth.

APPENDIX A

Findings from the PINTEC Database[1]

The initial and final periods of this cross-section analysis are 1997 and 2001, respectively. The econometric models divided firms into three categories: (a) firms that innovate and differentiate their products, (b) firms specialized in standard products, and (c) firms that do not differentiate their products and have lower productivity. One productivity measure used in the analysis was the log of potential value added per worker (log PVA per worker), measured as the log of total net sales less operational costs less total wages divided by the number of workers. Results of this exercise are presented in table A.1 and corroborate econometric findings discussed in chapter 5.

Table A.1. Impact of Innovation and Exports on Manufacturing Firm Productivity in Brazil Measured by Log of PVA per Worker, 2001

Variable	General model		Firms that innovate and differentiate products		Firms that specialize in standard products		Firms that do not differentiate products and have lower productivity	
	Coeff.	SE	Coeff.	SE	Coeff.	SE	Coeff.	SE
Constant	−17.50	−7.45	94.40	39.17	−35.50	11.20	−21.90	10.90
Product innovation (dummy)	0.23	0.06			0.35	0.06	0.03	0.10
R&D expenditures/total sales	0.20	0.06	0.15	0.14	0.48	0.07	0.04	0.17
(R&D expenditures/total sales)^2	0.01	0.00	0.00	0.00	0.02	0.00	0.00	0.00
Exporter (dummy)	1.61	0.11			0.47	0.14	0.00	0.00
Exports/total sales	0.13	0.01	0.04	0.03	0.07	0.01	−0.28	0.26
(Exports/total sales)^2	0.01	0.00	0.00	0.00	0.00	0.00	−0.08	0.01
Average schooling of workforce	0.63	0.05	0.96	0.18	1.29	0.08	0.10	0.08

(continued)

Table A.1. (*continued*)

Variable	General model		Firms that innovate and differentiate products		Firms that specialize in standard products		Firms that do not differentiate and have products lower productivity	
	Coeff.	SE	Coeff.	SE	Coeff.	SE	Coeff.	SE
Average experience of workforce	0.20	0.17	2.19	0.77	−0.36	0.26	0.59	0.26
Multinational (dummy)	0.50	0.05	0.47	0.09	0.39	0.05	0.30	0.21
Firms that innovate and differentiate products	0.63	0.05						
Firms that specialize in standard products	0.53	0.04						
R^2	0.60		0.59		0.50		0.56	
F	72.70		6.39		26.90		24.40	

Source: Arbache 2005.

Note: White's standard error (SE) estimates. Controlled for location (state), industrial sector, average age of workforce, marketing expenditures as a share of total sales, and labor turnover.

APPENDIX B

Econometric Analysis of the Relationship among R&D, Innovation, and Productivity Using ICS Data for Firm-Level Analysis

The econometric model consists of three equation sets that were estimated together, with results reported in table B.1 and table B.2.

First, the R&D equations model the sequential processes by which each firm determines its optimal level of investment in R&D. Using a Heckman selection model, the equations estimate the probability of a firm investing in R&D and also the firm's level of investment (the R&D intensity) once the investment decision is made.

Second, the innovation equations model the level of innovation, which is dependent on firm-specific characteristics and the R&D investment per employee. Two different innovation measures are considered: (a) an innovation dummy, which takes the value of 1 if the firm brought a new product to market or introduced a major new manufacturing process in the previous three years and (b) the innovation intensity, or the actual number of new products or processes developed by the firm in the same three-year span.

In the third set, the productivity equations are based on the standard Cobb-Douglas production function framework in which the observed value added per employee is dependent on labor inputs, capital inputs, and an unobserved productivity factor. The framework of Escribano and Guasch (2004) allows for use of observed investment climate variables as proxies for the last component.

Table B.1. Estimations for R&D, Innovation (Dummy), and Productivity

	Research equations	
	Selection equation	Log (R&D expenditures per worker)
Market share	0.178 [0.253]	1.276*** [0.460]
Diversification	−0.001 [0.002]	0.002 [0.005]
Professionals in labor force	4.510*** [0.847]	0.963 [1.189]
Overdraft	0.158* [0.095]	−0.181 [0.183]
Employment (log)	0.173*** [0.041]	−0.300*** [0.086]
		000.000
Constant	−1.291*** [0.250]	8.993*** [0.629]

Innovation equation	
Innovation dummy	
Log (R&D expenditures per worker)	0.193 [0.133]
Profit share reinvested	0.035 [0.083]
Professionals in labor force	1.893* [0.918]
Overdraft	0.116 [0.114]
Employment (log)	0.067 [0.056]
Constant	−1.820 [1.221]

Productivity equation	
Log (value added per worker)	
Innovation dummy	0.540*** [0.111]
Log (capital stock per worker)	0.276*** [0.029]
Log (inspections)	−0.013 [0.110]
Bribe tax	0.947 [1.033]
Share of workers using computers	1.274*** [0.260]
Constant	7.045*** [0.559]
Capacity utilization	0.629*** [0.206]
Publicly listed company	0.488** [0.223]
Quality certificate	0.291*** [0.093]

Professionals in labor force	0.016
	[0.956]
Overdraft	0.058
	[0.121]
Employment (log)	0.116**
	[0.048]
Log (power interruptions)	−0.063
	[0.059]
Losses due to transport interruptions	−3.422***
	[1.279]
Managerial time dealing with regulations	−3.334**
	[1.509]

Source: Correa et al. forthcoming.
Note: Optimal asymptotic least squares estimation. Robust standard errors in brackets. Regressions include 8 industry dummies, 12 state dummies, and a constant.
*Significant at the 10 percent level.
**Significant at the 5 percent level.
***Significant at the 1 percent level.

Table B.2. Estimations for R&D, Innovation (Intensity), and Productivity

	Research equations		Innovation equation		Productivity equation	
	Selection equation	Log (R&D expenditures per worker)		Innovation intensity		Log (value added per worker)
Market share	0.178	1.276***	Log (R&D expenditures per worker)	0.409*	Innovation intensity	0.480***
	[0.253]	[0.460]		[0.177]		[0.039]
Diversification	−0.001	0.002	Profit share reinvested	0.045	Log (capital stock per worker)	0.275***
	[0.002]	[0.005]		[0.110]		[0.029]
Professionals in labor force	4.510***	0.963	Professionals in labor force	−0.260	Log (inspections)	−0.008
	[0.847]	[1.189]		[0.710]		[0.110]
Overdraft	0.158*	−0.181	Overdraft	0.020	Bribe tax	0.939
	[0.095]	[0.183]		[0.165]		[1.033]
Employment (log)	0.173***	−0.300***	Employment (log)	0.277***	Share of workers using computers	1.233***
	[0.041]	[0.086]		[0.074]		[0.260]
Constant	−1.29***	8.993***	Constant	−3.290**	Capacity utilization	0.625***
	[0.250]	[0.629]		[1.622]		[0.206]
					Constant	6.853***
						[0.554]

Publicly listed company	0.480**
	[0.223]
Quality certificate	0.291***
	[0.093]
Professionals in labor force	1.098
	[0.715]
Overdraft	0.121
	[0.124]
Employment (log)	0.038
	[0.038]
Log (power interruptions)	−0.064
	[1.509]
Losses due to transport interruptions	−3.570***
	[1.279]
Managerial time dealing with regulations	−3.352**
	[0.059]

Source: Correa et al. forthcoming.
Note: Optimal asymptotic least squares estimation. Robust standard errors in brackets. Regressions include 8 industry dummies, 12 state dummies, and a constant.
*Significant at the 10 percent level.
**Significant at the 5 percent level.
***Significant at the 1 percent level.

APPENDIX C

Assessing Partial Effects of Firm Size Associated with Partial Effects in Explanatory Variables

In the Probit estimations, the dependent variables take the value of 1 if an effect is observed for a given firm; observations are pooled across firms. Tables C.1, C.2, and C.3 report the marginal effects of various variables, making it possible to assess the magnitude of the partial effects associated with changes in explanatory variables for each dependent variable.

As in the larger pooled datasets, the effects of size persist when controlled simultaneously with indicators for exporting, foreign ownership, and regional location in a regression framework. Small (20 to 99 workers), medium (100 to 499 workers), and large firms (500 plus workers) have higher (and increasing) probabilities of investing in R&D (9, 17, and 29 percent, respectively), getting an ISO certificate (11, 23, and 42 percent, respectively), providing worker training (20, 42, and 54 percent, respectively), and developing a new product (7, 9, and 16 percent, respectively) than micro firms (fewer than 20 workers). Results corroborate previous studies. For example, Mohnen and Dagenais (2002) found the propensity to innovate in Denmark to be significantly determined by firm size (that is, employment) and industrial sector. Lee (2004), studying the determinants of innovation among Malaysian manufacturers, found that larger firms were more likely to innovate than their smaller counterparts. De Negri (2006) also found size (natural logarithm of employment) to be a highly significant determinant of innovation probability by Brazilian firms.[1]

Exporters and firms with some foreign ownership also show higher levels of innovative activities, even when controlled for size, sector, and region. Exporters are significantly more likely to invest in R&D (12 percent), acquire an ISO certificate (13 percent), provide worker training (10 percent), and establish joint ventures (3 percent) than nonexporters of the same size, sector, and region. Salomon and Shaver (2005), examining product innovation and patent application counts of a representative sample of Spanish manufacturing firms from 1990 to 1997, also found exporting to be positively associated with innovation. In addition, firms with foreign capital present higher probabilities of having an ISO certificate (29 percent), providing

Table C.1. Marginal Effects on Innovation Inputs and Outputs in Brazil

Independent variables	R&D	ISO	Worker training	Joint venture	Technical licenses	New product	Improved line
20 to 99 workers	0.085*	0.113***	0.200***	0.004	0.017	0.070**	0.029**
	[2.61]	[3.80]	[5.61]	[0.38]	[1.11]	[2.37]	[2.42]
100 to 499 workers	0.172***	0.234***	0.420***	0.029**	0.087***	0.090**	0.027**
	[4.29]	[5.67]	[9.58]	[2.16]	[3.75]	[2.51]	[2.01]
500-plus workers	0.293***	0.425***	0.535***	0.034*	0.303***	0.160***	0.022
	[4.19]	[5.68]	[7.31]	[1.61]	[5.45]	[2.64]	[0.96]
Exporter	0.117***	0.130***	0.103***	0.026**	0.006	0.043	0.019
	[3.72]	[5.72]	[3.45]	[3.10]	[0.50]	[1.45]	[1.50]
Foreign ownership	0.001	0.292***	0.252***	0.052***	0.263***	−0.010	−0.013
	[0.01]	[5.40]	[3.83]	[3.48]	[7.47]	[0.17]	[0.50]
Observations	1,642	1,562	1,639	1,640	1,640	1,640	1,640
LR χ^2 (d.f.=16)	120.64	521.26	413.86	111.96	203.73	70.31	46.49
Pseudo R^2	0.053	0.342	0.196	0.198	0.233	0.034	0.067

Source: ICS-Brazil.
Notes: Z-value is in brackets. For brevity, sector and regional variables were not included in table C.1. The wood and furniture sector is the omitted category for sector. Southeast is the omitted category for region. Micro is the omitted variable for size.
*Significant at the 10 percent level.
**Significant at the 5 percent level.
***Significant at the 1 percent level.

Table C.2. Marginal Effects on Innovation Inputs and Outputs in Brazil

Independent variables	R&D	ISO	Worker training	Joint venture	Technical licenses	New product	Improved line
Exporter	0.093***	0.134***	0.094**	0.025***	0.007	0.051	0.021
	[2.66]	[5.43]	[2.84]	[2.90]	[0.53]	[1.56]	[1.45]
Sales to exporter/total sales	0.035	−0.054**	−0.303	0.044	−0.010	−0.060	−0.011
	[0.66]	[2.04]	[0.62]	[0.42]	[0.60]	[1.17]	[0.41]
Foreign ownership	−0.025	0.233***	0.210***	−0.004***	0.243***	−0.033	−0.015
	[0.41]	[4.47]	[3.16]	[3.12]	[6.94]	[0.55]	[0.57]
Sales to foreign firms/ total sales	0.097***	0.099***	0.124***	0.009	0.016	0.052*	0.003
	[3.18]	[4.65]	[4.24]	[1.31]	[1.34]	[1.83]	[0.24]
Observations	1,642	1,562	1,639	1,640	1,640	1,640	1,640
LR χ^2 (d.f.=18)	131.28	546.31	432.02	113.81	205.87	74.90	46.70
Pseudo R^2	0.058	0.358	0.205	0.201	0.236	0.036	0.067

Source: ICS-Brazil.
Notes: Z-value is in brackets. For brevity, sector, regional, and size variables were not included in table C.2. The wood and furniture sector is the omitted category for sector. Southeast is the omitted category for region. Micro is the omitted variable for size.
*Significant at the 10 percent level.
**Significant at the 5 percent level.
***Significant at the 1 percent level.

Table C.3. Marginal Effects on Innovation Inputs and Outputs in Brazil

Independent variables	R&D	ISO	Worker training	Joint venture	Technical licenses	New product	Improved line
Employees with high	0.001**	0.001**	0.002***	−0.001*	0.001*	0.002**	0.001***
school (%)	[1.96]	[2.17]	[4.11]	[1.70]	[1.64]	[2.52]	[3.18]
Employees with some	0.008***	0.004***	0.006***	−0.001	0.001	0.004***	0.001**
college (%)	[5.13]	[4.14]	[4.14]	[0.18]	[0.73]	[2.68]	[1.99]
Loan	0.011	0.002	0.021	−0.002	−0.005	0.012	0.002
	[0.39]	[0.09]	[0.80]	[0.36]	[0.49]	[0.49]	[0.16]
Observations	1,631	1,554	1,630	1,631	1,631	1,631	1,631
LR χ^2 (d.f. = 21)	161.83	566.68	462.60	117.13	207.69	91.06	63.80
Pseudo R^2	0.072	0.375	0.221	0.210	0.240	0.044	0.092

Source: ICS-Brazil.

Notes: Z-value is in brackets. For brevity, variables for sector, region, size, export status, foreign ownership, the share of sales to exporters, and the share of sales to foreign-owned firms were not included in table C.3. The wood and furniture sector is the omitted category for sector. Southeast is the omitted category for region. Micro is the omitted variable for size.
*Significant at the 10 percent level.
**Significant at the 5 percent level.
***Significant at the 1 percent level.

worker training (25 percent), engaging in joint ventures (5 percent), and holding technology licenses (26 percent) than domestically owned firms. Two separate studies of Scottish and German manufacturing firms, respectively, show a significant and positive relationship between foreign ownership and innovation (Bertschek 1995; Love et al. 1996). Lofts and Loundes (2000), using a sample of Australian firms between 1994 and 1997, also found foreign shareholding to be a determinant of innovative activity levels in Australia.

APPENDIX D

The Primary and Secondary Education Systems

Institutional Arrangements for Basic Education

In the Constitution of 1934 the Brazilian government defined education as a basic right for all its citizens. Today, Brazil's basic education system is divided into (a) preschool; (b) the *ensino fundamental*, an eight-year cycle joining the former primary (*primário*) and lower-secondary (*ginásio*) levels; and (c) a three-year "intermediate" cycle (*ensino médio*). Preschool education covers the social development of children through age six. *Ensino fundamental* (for 7–14 year olds) is divided into two stages (grades 1–4 and 5–8), with national testing conducted at the end of each stage and an increasingly diversified curriculum and instructional organization during the second half of the cycle. *Ensino médio* consists of grades 9–11 and is intended for students aged 15–17. The National Education Law—LDB (*Lei de Diretrizes Basicas*)—describes *ensino médio* as the "final phase of basic education" to which all citizens are guaranteed access.

Despite the formal unification of grades 1–4 and 5–8 into a continuous *ensino fundamental* cycle, grades 5–8 are a complement to *ensino médio*. The clearest evidence of this is school organization: most schools that offer grades 9–11 also offer grades 5–8, as indicated in table D.1.

Prior to the 1988 Constitution, all three levels of government (municipal, state, and federal) were involved in the financing and provision of all levels of education. The uncoordinated coexistence of federal, state, and municipal education systems for decades has been a primary source of inequity and inefficiency within Brazilian basic education. Building upon new guidelines in the 1988 Constitution, the 1996 LDB further delineated administrative responsibilities as follows: municipal and state governments share responsibility for financing and provision of grades 1–8, while state governments are primarily responsible for the provision of grades 9–11.

State systems in Brazil currently enroll 23 percent of grade 1–4 students, while municipal systems enroll 67 percent. The picture for lower secondary is quite different: state systems account for nearly 53 percent of students, while municipal systems account for 37 percent. States also bear the lion's share of

Table D.1. Structure of Brazil's Basic Education System

Age	Grade	Brazilian structure	Report terminology
7	1st	Ensino fundamental	
8	2nd	(Series 1st–4th)	
9	3rd		
10	**4th**		→ Primary
11	5th	Ensino fundamental	
12	6th	(Series 5th–8th)	
13	7th		→ Lower secondary
14	**8th**		
15	1st	Ensino médio	
16	2nd		→ Upper secondary
17	**3rd**		

Source: Authors.

Note: Bold indicates grade at which SAEB, the National Achievement Test, is applied.

upper-secondary education with 85 percent of enrollments, while municipalities account for nearly 2 percent, the private sector for nearly 10 percent, and the federal government for less than 1 percent. Most state governments are transferring the school administration of grades 1–4 to municipalities.

Together, primary and secondary enrollments account for some 42.5 million students, (33.5 million in primary and approximately 9 million in upper secondary). According to the 2003 teacher census, there are more than 1.5 million teachers, 34 percent of whom lack university education. Of all basic education teachers, 39 percent are hired by the states, 48 percent by the municipalities, and 12 percent by the private sector.

Coverage and Access—Consolidating Gains, Addressing New Challenges

There is little doubt that Brazil has made significant progress in expanding access to all educational levels in recent years. Table D.2 provides a quick snapshot of the gains. For primary enrollment in grades 1–8, the country can now boast near-universal coverage, with much of this improvement occurring in aggregate terms before 1999. When the focus shifts to specific target groups—such as the rural poor in the Northeast—the gains continue to be dramatic. Evidence for preprimary education shows steady improvement between 1999 and 2003. Finally, the immense increase in secondary enrollment rates—from 15 percent nationally in 1990 to 76 percent in 2003—is clearly the most important development in coverage and access in recent years.

Table D.2 also highlights some of the remaining educational challenges. Preprimary access is expanding slowly, but more work is needed to improve coverage and quality. Efficiency issues also must be addressed (see the next section, "The Policy Imperative of Improving School Quality" for fuller discussion).

Table D.2. Coverage, Access, and Efficiency, 1990–2003

percent

Indicator	Year			
	1990	1999	2001	2003
Preprimary net enrollment	—	44	50	53
Primary net enrollment	86	91	94	93
Secondary net enrollment	15	66	71	76
Over-age enrollment (total)	—	56	48	38
Repetition rates				
Grade 1	—	31	29	—
Grade 4	—	14	13	—
Secondary	—	18	18	19
Survival to last grade of primary	—	80	84	—
Transition from lower to secondary	—	84	84	—

Source: UNESCO Education Statistics 2006; UNDP Human Development Report 2005.
Note: — = not available.

At the primary level, repetition rates are decreasing steadily but remain high by any standard. At the secondary level, the challenges are more pronounced, as shown by the increase from 18 percent to 19 percent in repetition rates between 2001 and 2003. As coverage expands to include social sectors that were largely excluded from the system, outcomes like average achievement and efficiency are unlikely to improve or only will improve slowly. Of course this apparent tradeoff between quantity and quality is not a given, and policy makers can act to bring about more dynamic quantitative and qualitative improvements. Prospects for doing so, however, should be tempered by awareness that simultaneous improvements in coverage and quality have proven difficult to realize in most countries.

The improvements in educational coverage and access represent a major public policy success in Brazil. Some important antecedents in the history of educational policy making in Brazil have already been reviewed, as well as in Rodríguez and Herrán (2000). Here, we only touch on the more salient points that help explain the data. As previously mentioned, the 1996 LDB legally demarcated the roles of federal, state, and municipal governments, but it also mandated minimum standards for schools. Minimum standards were given some teeth via financial support through the FUNDEF (Fund for the Development of Fundamental Education & Valorization of Teachers, also known as FVM) program, which requires state and municipal governments to devote a certain percentage of revenues to basic education. The funds are then distributed based on enrollments, which doubtlessly helped expand access and coverage in fundamental education in Brazil. FUNDEF's accomplishments are discussed in detail in several recent Bank studies. However, as Rodríguez and Herrán (2000) note, the FUNDEF scheme also contributed to a bottleneck of basic education graduates who were unable to continue their studies. This has, in fact, contributed to creation of the FUNDEB (Fund for the Development

Figure D.1. Average Years of Schooling for Brazilians 15 and Older in Comparison with Other Selected Countries, 1960 and 2000

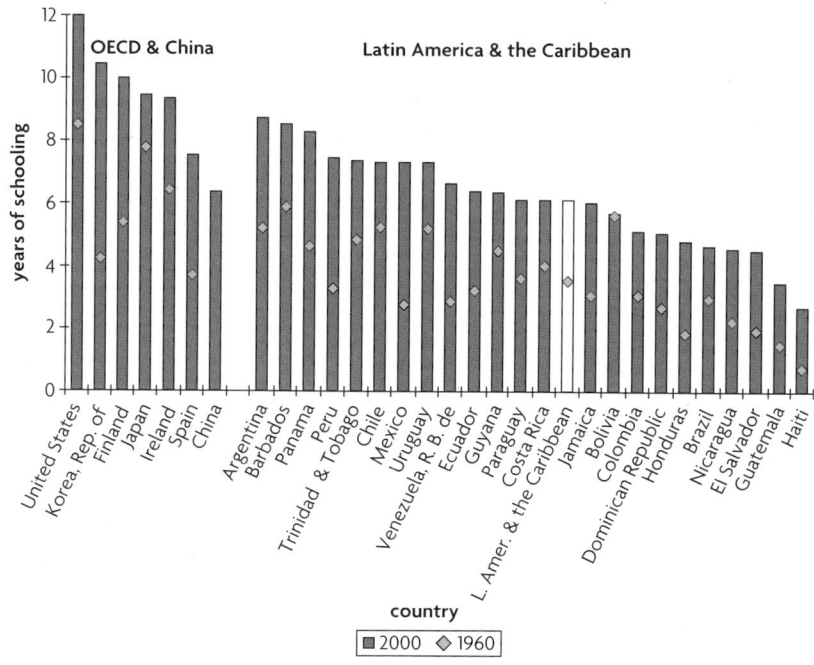

Source: IDB 2006.

Figure D.2. Net Enrollment Rate for Primary Education, 1990 and 2002

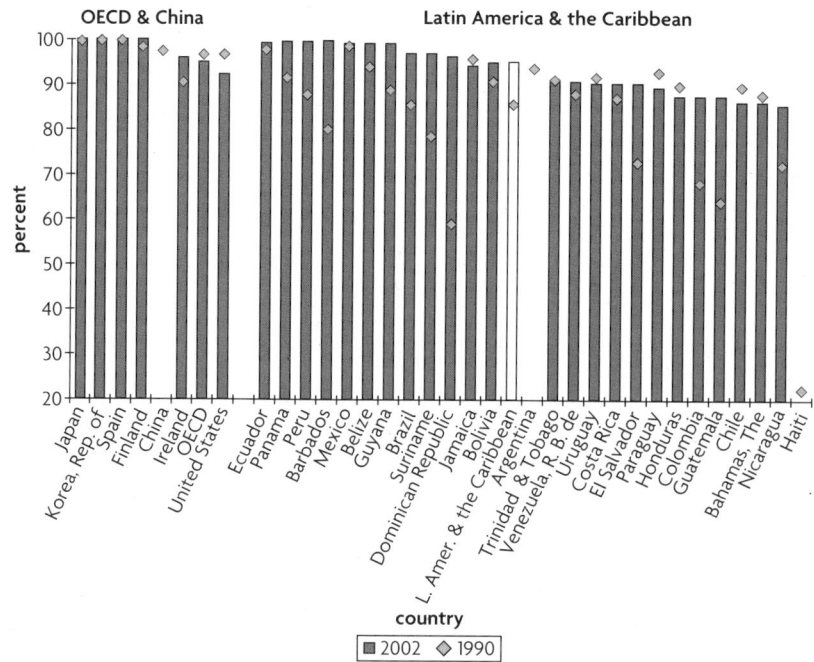

Source: IDB 2006.

Fgure D.3. Net Enrollment Rate for Secondary Education, 1990 and 2002

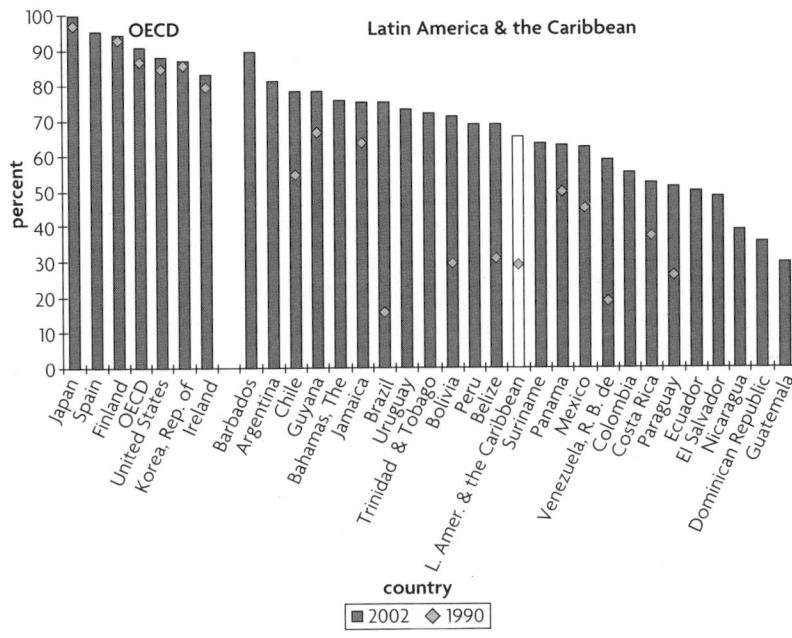

Source: IDB 2006.

Figure D.4. Tertiary Enrollments for Brazil vs. OECD Comparators: NERs, 1991–2003

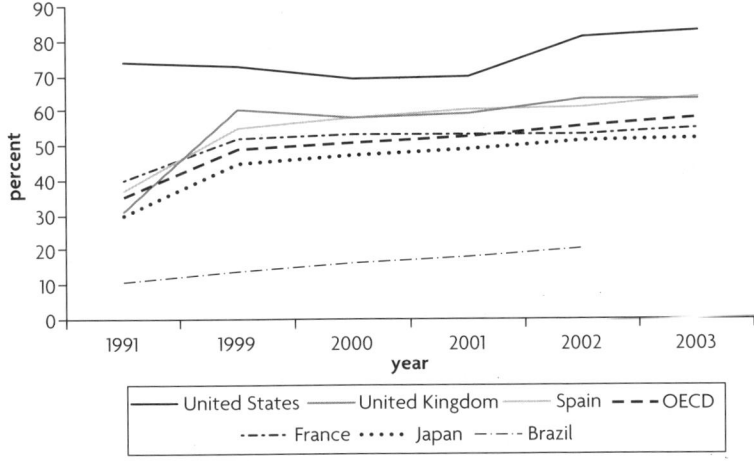

Source: KAM 2006.

of Secondary Education) program, a funding scheme that includes coverage of both preprimary and upper-secondary education levels to better coordinate supply and demand.

Institutional developments related to the LDB and FUNDEF have been complemented by a host of policies to address additional demand and supply issues. The *Bolsa Escola* and *Bolsa Família* programs are probably the best-known governmental efforts to stimulate educational demand through direct subsidization of attendance. The programs have become so popular that they now play a central role in the federal government's social protection policy.

In addition to school finance reform and demand-side initiatives like *Bolsa Família*, the government also has helped institute changes within schools. Some of those changes have been positive side effects of the general expansion of education in the 1990s and the creation of minimum standards, which helped push up teacher preparation levels. Other attempts focused on bringing change through decentralized decision making, as in the PDDE (Projeto Dinheiro Direito na Escola) Direct School Funding Project and *Plano de Desenvolvimento da Educação* (Plan for Educational Development, PDE) initiatives. Both programs provide direct funding to schools, which are then responsible for carrying out school-defined priorities. In addition to developing local capacity, these programs try to reduce educational inequalities within and between municipalities. And there is some evidence that they result in higher retention rates and increased efficiency (Carnoy et al. 2008).

In sum, exogenous forces were not responsible for the notable changes in coverage in Brazilian basic and secondary education during the past 15 years: rather a specific public policy goal was defined and achieved. This point is important because it reinforces awareness that the government has a key role to play and that Brazil, despite being a developing country, is not without resources to address social problems. The positive consequences of these actions have the potential to create a virtuous circle because the ratcheting up of human capital levels not only means that current cohorts of young people have more skills and opportunities than their parents and (especially) grandparents had, but their children in turn should also benefit from having better-educated mothers and fathers.

Nevertheless, the extent to which quantitative educational changes profoundly affect individuals and society depends greatly on quality. In other words, increased access cannot be judged solely by generating more credentials and meeting targets like Education for All (EFA). Expanding coverage is a means for increasing relevant skills that individuals can use to improve their lives. Of course the educational system is not solely responsible for how these skills are generated, let alone how they perform on the labor market. For all these reasons, coverage indicators have limits as measures of human capital levels.

Turning from quantity to quality spotlights the challenges that remain in Brazilian education, some of which are reflected in the repetition rates in table D.2. The following subsections will discuss these issues in terms of coverage, quality, and equity. The guiding theme is the need to redouble efforts to guarantee that primary and secondary education graduates enter the workforce—or the university—with the kinds of basic skills needed for success.

The Policy Imperative of Improving School Quality

Few topics receive more attention in education policy and research circles than school quality. This is true in industrialized and developing countries alike, highlighting the need for all school systems to be continually vigilant amid growing concerns about global competitiveness. How is school quality best measured? Test scores or graduation rates are commonly used indicators of school system performance. However, from a policy-making standpoint, the inputs for creating outcomes like student achievement and retention are more important. These include the school climate, the work of the school director, and the teaching and learning environment inside the classroom. Measuring these inputs is not easy, making it difficult to use them to comparatively assess quality. It also complicates researchers' attempts to identify these processes as determinants of educational outcomes.

A second conceptual complication involves who will decide what quality means. In a very narrow economic sense, the labor market decides what school quality is, based on how different credentials predict earnings; but only in an idealized world do the skills learned in school perfectly track a person's earnings. School systems also can monitor quality by creating minimum standards for what schools should look like, or by using standardized tests to monitor school performance. Finally, analysis also must allow for individual families to decide what quality is, especially in countries where the state does not actively enforce school attendance laws. For example, when a child is pulled from school because the family does not think the child is learning, or doesn't believe the school experience is valuable, then the family's definition of school quality takes on added significance (Marshall, in press).

This discussion of the complexities involved is not meant to suggest that school quality is immeasurable—it is constantly being assessed by someone. But the limits of simplistic formulations of school quality based solely on one kind of input or output must be kept in mind. Indeed, a range of measurements must be taken into account when building an empirical profile of quality in a country like Brazil. In the following sections we do just that, focusing on several elements: (a) education spending; (b) standardized test scores and pass rates; (c) the teaching and learning environment in schools and classrooms, including how students are taught and what they are taught (i.e., curriculum); and (d) the accountability system.

Education Spending

We have already reviewed the institutional arrangements for education provision in Brazil. Now we turn to the specifics of education spending and how institutional structures help determine resource allocation. Based on recent experiences, several points are clear. First, the focused efforts on local financing sources (states and municipalities), combined with minimum spending guarantees (through FUNDEF), played a major role in expanding the reach of basic education (grades 1–8). Second, education financing relies on capitation, that is, transfers of financial resources are based on the number of students being served. While this is a fairly natural criterion, it has rarely been used in most countries in Latin America, where education financing "follows the teacher" because coverage is largely confined to teacher salaries and teachers are hard to relocate based on rapid demographic changes in student population. Third, the funding success at the basic level not only has gone unreplicated at other levels, but the focus on primary education (grades 1–8) has come at the expense (to some degree) of spending on preprimary and secondary education. This is less a criticism of FUNDEF than a recognition of how education priorities may evolve over time. Fortunately, FUNDEB was recently established to extend FUNDEF's financing success to other education levels, although exactly how the new program will operate is still under discussion. Fourth, despite steady improvement in education funding, Brazil still lags behind its neighbors and (especially) developed countries in spending per student.

This last issue is especially important moving forward, and is also potentially the most controversial. The analysis by Abrahão (2005) of financing in Brazil shows that education spending has increased from 3.9 percent of GDP in 1995 to roughly 4.3 percent in 2002. In real terms, this is roughly a 10 percent increase in a fairly short time span. But as table D.3 and figure D.5 show, Brazilian spending per student is low by international and even regional standards.

The comparatively low levels of per pupil spending in Brazil have serious consequences for quality and equity. As Abrahão (2005) notes, substantial supporting evidence comes from the Programme for International Student

Table D.3. Comparative per Student Spending by Education Level
U.S. dollars

Country	Education Level		
	Preprimary	Primary	Secondary
Brazil	*1,044*	*832*	*864*
Latin America			
Argentina	1,745	1,655	2,306
Chile	1,766	2,110	2,085
Mexico	1,410	1,357	1,915
Paraguay	—	802	1,373
Peru	339	431	534
OECD countries			
Denmark	4,542	7,372	8,113
France	4,323	4,777	8,107
Germany	4,956	4,237	6,620
Korea, Rep. of	1,913	3,714	5,159
United States	8,522	7,360	8,779

Source: Abrahão 2005: table 5.

Note: All numbers refer to PPP adjusted per-student dollar expenditures/year. — = not available.

Figure D.5. Comparative Public Spending on Education as a Percentage of GDP, 1990 and 2002

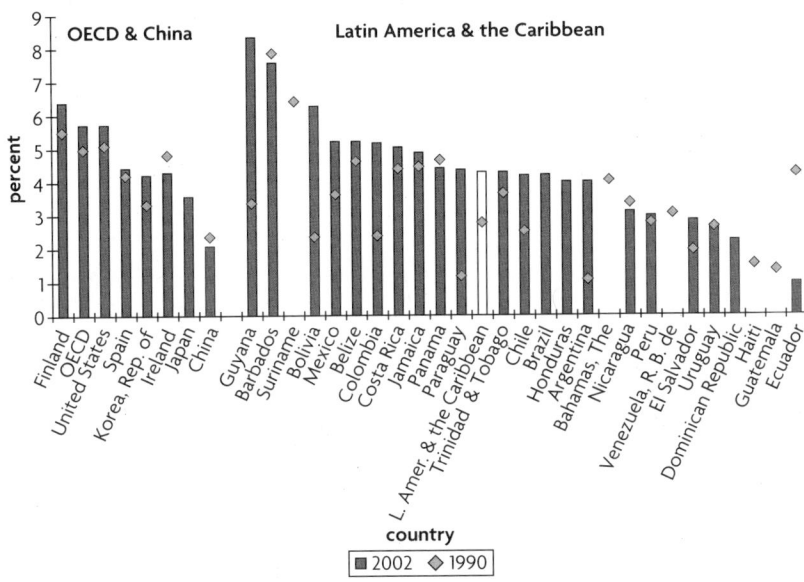

Source: IDB 2006.

Note: Data for the United States, Japan, Brazil, and Ecuador are for 2001. Data for China and Honduras are for 1999 and 1998; respectively.

Assessment (PISA) testing project in 2003, which shows a fairly strong correlation between spending and performance on the tests (the PISA data will be examined more closely below). The fact that Brazil lags behind industrialized countries in spending is unsurprising. Yet the discrepancies are notable when compared with Brazil's closest neighbors and with competitors such as Argentina, the Republic of Korea, and Mexico.

Macroanalyses of spending are useful for providing a general overview, but outcomes such as spending per pupil result from a very complicated interplay between politics, economics, and other factors. Any discussion of present or future education financing in Brazil must grapple with the historical realities of institutionalized inequality. This does not mean that structure is supremely important or that policy makers are helpless to redress massive inequalities; the country's experiences with basic education in the past 15 years strongly belie such assertions. Yet the issue cannot be reduced to a simple question of policy choices, and one must be realistic about what pace of change is feasible.

Despite their utility as benchmarks for the government's commitment (or ability) to finance education, the kinds of internationally comparable data shown in table D.4 have definite limitations. For example, according to table D.3 the United States is the biggest spender in education, and Korea ranks in the middle (or lower). And yet Korean students consistently score among the highest on international mathematics exams (for example, the Trends in International Mathematics and Science Study [TIMSS] and PISA), while

Table D.4. Breakdown of Basic and Secondary Education Spending by Brazil and Four Comparators

percent of total education spending

Country	Spending type		
	Salaries	Other current	Capital
Brazil	75	16	9
Argentina	89	10	1
Mexico	92	5	3
Korea, Rep. of	59	23	18
Malaysia	49	11	40

Source: UNESCO Institute for Statistics 2006.
Note: All numbers refer to percentage of total budget at education ISCED (International Standard Classification of Education) levels 1–4.

U.S. students are in the bottom tier of the distribution. Clearly there are choices about how to spend public resources, and based on evidence from international tests, some countries appear to be more proficient at maximizing their results. This is the promise that good policy making holds out, in theory, for poorer countries to catch up with their more developed neighbors. And this provides a very useful segue into the issue of the makeup of spending, which is arguably of equal or greater importance than the overall level of spending.

Table D.4 provides a basic overview of Brazil's spending structure compared with a handful of other countries. The results show the tendency in Latin America to focus expenditures on salaries, whereas the two sample countries from Asia devote a larger percentage of their budgets to other current and capital outlays. The important point is that countries like Brazil have little left to invest in quality upgrades because they are busy hiring teachers to keep pace with burgeoning student populations.

Are teachers overpaid in Brazil? This question has received much generic attention, especially in countries where teachers' unions are active. According to UNESCO data, Brazil's ratio of primary and secondary education teacher salaries to spending per student is one of the highest in the world and is nearly three times as high as the OECD ratio (Di Gropello 2006). Given the performance of Brazilian students on international tests (summarized below), this would appear to be an inefficient use of resources because teachers are paid as much as or more than teachers in countries that have higher achievement. One factor that exacerbates the challenge is the generous pension system and the rewards structure for Brazilian public employees.

The makeup of educational expenditures could be the focus for a study of its own, and the topic of teacher labor markets also looms large. The evidence clearly shows that Brazil is not spending as much on education as needed to compete internationally. Nevertheless, this does not justify automatically ratcheting up expenditures along traditional lines such as hiring more teachers to reduce class sizes or raising pay. In fact, some evidence suggests that

Brazilian teachers receive a disproportionate share of resources, or at least are not performing to the level their pay would predict.

There is one final component to Brazil's spending dilemma: repetition. In an unpublished policy note commissioned by the World Bank, Ioschpe (2006) estimates that students repeating primary and lower secondary grades costs Brazil R\$12.6 billion annually. This represents a significant loss of budget resources, although how much depends to some degree on how much learning takes place. Nevertheless, Brazil's high rates of repetition (detailed below) continue to put fiscal pressure on a system that is already stretched thin. Addressing this problem through more effective teaching and learning environments will not only accelerate human capital formation, but will thereby generate new resources to invest in improved student achievement.

Standardized Test Scores and Pass Rates: Low Efficiency Suggests Low Quality

Several references already have been made in this appendix to the low quality of public education in Brazil. Where does this belief arise? Criticisms commonly refer to results from the SAEB national testing system and from international tests (such as PISA). Student scores, whether considered domestically or compared internationally, are very low. The SAEB results are based on tests constructed by Brazilian curriculum experts and are designed to communicate student results in words—through levels of performance—rather than as simple statistical summaries. The 2003 SAEB shows that student abilities in grades four and eight are far below what is expected based on the intended curriculum. For example, the average of 177.1 in grade four mathematics is significantly under the 200-point level that is considered an acceptable level of knowledge. Students who scored at the 2003 average level of proficiency can only do basic multiplication and tell time with digital clocks instead of traditional timepieces. The language proficiency results in Portuguese and for grade eight are also considerably below expected achievement levels.

If expectations are not being met, is the direction of SAEB results at least improving? When the SAEB time series was analyzed rigorously, Biondi (2007) found several trends. A slight but statistically significant improvement that started in 2001 has been noted in fourth-grade students in both mathematics and Portuguese. Previously—starting in 1995—fourth-grade performance had been worsening. This switch is unsurprising, since universal enrollment of 1st–4th graders occurred in the mid-1990s, so that the poorest students with the lowest social capital finally were being schooled and tested. Once the impact of serving a massive inflow of underprivileged students was absorbed by the system, average performance stabilized and began to inch upward. The wave of new students from the mid-1990s is now old enough to affect results from the 8th and 11th grades, helping explain the downward trend observed in math and Portuguese performance for those grades between 2003 and 2005. However, intertemporal comparisons of student knowledge levels in a country the size of Brazil are complicated by numerous factors, not the least of which is the fact that more students are remaining in the system. Although more time is needed to establish the exact learning trend countrywide, it is already apparent that the overall level of knowledge is low, and great improvement is required to raise Brazilian achievement levels to those being posted by key neighbors and international competitors.

Brazil's participation in the 2000 and 2003 PISA international achievement study provides still more dramatic evidence of the work that remains in improving quality. In both years Brazilian eighth-grade students scored at the bottom of the distribution in mathematics, below countries such as Indonesia and Mexico and far below high scorers like Korea. In terms of proficiency, the results showed that more than half of Brazilian students fell below even Level 1 on a six-level ascending scale (1–6). In other words, the PISA results largely confirm the low proficiency attainments demonstrated by SAEB, but on an international scale.

The dramatic increase in Brazil's matriculation coverage in recent years affects these results in several ways. First, per pupil expenditures on education are lower, even compared with some other Latin American countries. A cohort effect has also emerged because more students of low socioeconomic status are remaining in school longer, requiring resources to be diverted into hiring new teachers to keep up with the advancing wave. Test score comparisons underline the obvious importance of improving quality and spotlight the dangers of relying on coverage indicators to measure the "health" of Brazilian education. Test score data also serve an important monitoring function, which is why the high-quality work of the SAEB must continue. The same is true for participation in international testing, however painful the findings may be.

Table D.5 presents international data on grade repetition rates. Results show that, despite recent improvements in its internal efficiency, Brazil still has some of the highest repetition rates in the world. The consequences are easily detailed. First, equity is a serious concern because the poorest students tend to repeat more often and eventually leave school with fewer of the skills needed to rise out of poverty. Additionally, overall spending is affected, as previously discussed.

Why are Brazil's repetition rates so high? It makes intuitive sense that low school quality leads to high repetition and dropout rates. But we must be wary of concluding that low achievement is the sole proximate cause. Gomes-Neto and Hanushek (1994) show that grade repeaters score higher than nonrepeaters in Brazil, which suggests that other causative factors may be involved. Marshall's (2003b) analysis of repetition in Honduras reaches a similar conclusion. Several factors may be in play. Poorly trained teachers may use grade failure (or the threat of failure) to control student behavior, especially for older children. A stigmatizing effect also may be at work, whereby students are labeled as repeaters and treated differently as a result. Students may drop out of the system because of poor learning environments (fights or hazing, for example) or from boredom. In sum, we should be concerned about the potential for low achievement to reduce efficiency and overall attainment.

Table D.5. Repetition Rates in Brazil and Comparator Countries

percent

Country	Grade level			
	1	**2**	**3**	**6**
Brazil	*28*	*19*	*15*	—
Argentina	10	7	6	4
Chile	1	3	1	2
Guatemala	28	14	11	2
Mexico	8	8	5	0
Paraguay	14	10	7	0
Peru	6	14	11	3
Cambodia	18	11	8	2
India	4	3	4	—
Philippines	5	2	2	0
Vietnam	5	3	2	—
Ethiopia	19	14	13	—
Ghana	9	6	5	4
Kenya	6	7	6	6
Mozambique	26	25	25	24
South Africa	7	5	6	5

Source: UNESCO Education Statistics 2006.
Note: All numbers refer to repetition rates for a specific grade. Most of the data comes from the 2002 school year or from 2003 if 2002 is unavailable. — = not available.

Nevertheless, each outcome is a product of multiple factors, so simply raising overall achievement will not necessarily solve either problem.

The Teaching and Learning Environment in Schools and Classrooms

Repeated references have been made to deficient teaching and learning environments or, more generally, to low-quality schools. Such judgments are easy to defend based on outcome measures, especially from standardized testing. But improved policy making to redress those results requires deeper understanding of the specific mechanisms curtailing school quality.

Multiple sources of information are available to shed light on this concern. Qualitative studies of Brazilian classrooms and teachers are abundant as are quantitative studies of student test scores from SAEB (Barros and Mendonça 2001) and other test applications (Carnoy et al. 2008; Fuller et al. 1999). This report relies on data sources from international studies of Brazilian education, including the aforementioned PISA study from 2003 and the qualitative classroom comparisons conducted by Carnoy, Gove, and Marshall (2007). These results are extensive and cover multiple dimensions, and each represents a possible policy mechanism for improving school quality in Brazil.

Table D.6 briefly summarizes performance on the PISA 2003 test by Brazil and four other countries: Korea, Thailand, Mexico, and Uruguay. The list of "competitors" was chosen to create a somewhat diverse set of comparisons involving high- and medium-performing Asian countries as well as other Latin American countries. The Korean case is clearly the most important in terms of drawing lessons, based both on its rapid rate of development in the past four decades and its very high test scores. But Brazilian performance vis-à-vis the other countries is also likely to uncover clues about possible policies for improving Brazilian education and, by extension, competitiveness.

Figures D.6 through D.8 provide a somewhat broader picture of PISA results.

In all three subjects covered in table D.6, Brazilian upper-primary and lower-secondary students scored significantly below test takers in almost all other sample countries. For mathematics, the differences are especially large, while for language and (to a lesser extent) science, the gaps with Thailand, Mexico, and Uruguay are less large. In the case of Korea, the gap is very large (upwards of two standard deviations).

Examining the socioeconomic profile of test takers, we see that both Korea and Uruguay have more affluent students in their samples. But this is not

Table D.6. Overview of PISA Performance in Brazil and Four Other Countries, 2003

	Country				
Variable	Brazil	Korea, Rep. of	Thailand	Mexico	Uruguay
Mathematics score	356	542*	417*	385*	422*
Reading score	404	535*	420*	400	434*
Science score	392	539*	429*	404*	438*
Poverty index	−0.95	−0.10*	−1.18*	−1.12*	−0.35*
Parent education (years)	10.7	12.5*	8.9*	9.6*	12.2*
Marginal difference					
Math regression 1	—	120.4*	20.5*	6.9	25.6*
Math regression 2	—	104.4*	25.2*	13.6*	16.5*
Reading regression 1	—	71.3*	−24.9*	−26.6*	−7.4
Reading regression 2	—	62.0*	−20.2*	−17.4*	−14.3*
Science regression 1	—	90.3*	2.2	−6.6	12.3*
Science regression 2	—	77.1*	9.6*	−1.7	4.1

Source: PISA 2003.

Note: Sample weights are used for calculating means. T-test comparisons are based on individual comparisons between Brazil and each country separately. The poverty index is a standardized measure that is based on all participating countries, not just these five. Regression 1 includes basic controls for the student's grade, type of school and location, and the country dummies only. Regression 2 adds student and family SES background measures. Coefficients for regressions 1 and 2 refer to the marginal difference in achievement between each country and the excluded category, Brazil. Dashes are used to signify that data are not available.

*Difference in mean (or regression coefficient) is significant at 0.05 level.

Figure D.6. Quality of Education in Terms of Learning Outcomes from PISA, 2003

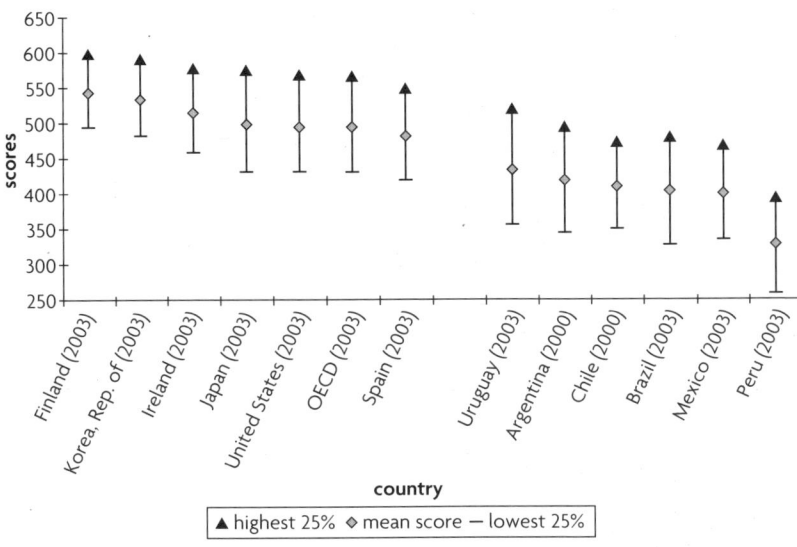

Source: IDB 2006.
Note: To date, only six Latin American countries have participated in PISA.

Figure D.7. Percentage of Students at Each Level of Proficiency on PISA Reading Scale, 2003

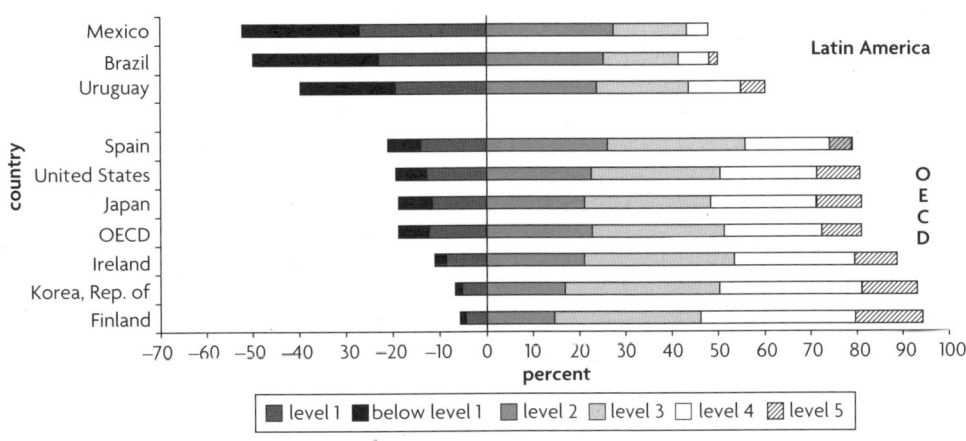

Source: IDB 2006.
Note: The zero-line constitutes a threshold below which performance is so low that even the most routine or obvious tasks are completed with difficulty.

so for Thailand and Mexico, which suggests that both countries are getting more from their education systems than Brazil is from its. This proposition is tested more directly at the bottom of table D.7. For each subject examined by PISA, a baseline regression is run that includes only basic controls for grade level, type of school, location, and student gender. Then controls are

Figure D.8. PISA Math/Space and Shape Scale, 2003

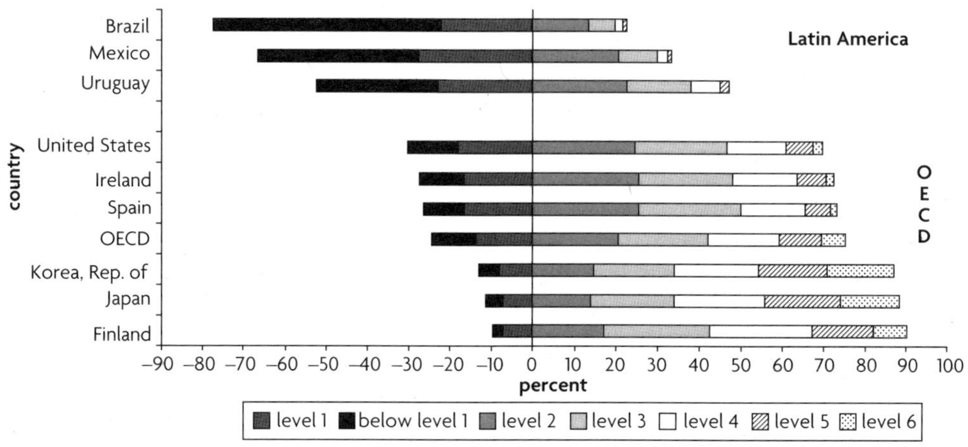

Source: IBD 2006.

Table D.7. School Resources, PISA 2003

	Country				
Variable	Brazil	Korea, Rep. of	Thailand	Mexico	Uruguay
Student-teacher ratio	33.6	16.3*	22.6*	—	16.8*
Math minutes/week	206	245*	227*	245*	189*
Total minutes/week	1,143	1,815*	1,836*	1,489*	1,345*
Shortages of					
Textbooks	2.41	1.29	2.42	2.19	2.86
Computers	2.92	1.55	2.73	2.62	2.96
Lab equipment	3.12	1.92	2.56	2.69	2.64
Computers per student	0.02	0.28*	0.05*	0.09*	0.06*
Teachers w/masters (%)	21.0	93.3*	96.2*	—	7.1*

Source: PISA 2003 database.
Note: — = No data available. Shortages are measured on a scale from 1–4, as reported by school director. 1 = Not at all, 2 = Very little, 3 = To some extent, 4 = A lot.
*Variable with a statistically significant impact.

added for socioeconomic status of the family in the second regression. Three results stand out. First, school quality is clearly superior in Korea compared with Brazil (and other countries), even if no specific information is identified yet on what accounts for such a big difference. Second, Brazilian performance vis-à-vis this group of countries is especially low in mathematics and, to a lesser extent, science. This is troubling given the importance of these skills in the larger picture of competitiveness and innovation. Third, the marginally more efficient Brazilian achievement in reading skills versus

Thailand, Mexico, and Uruguay must be noted because it suggests that favorable factors exist inside Brazilian schools to produce this outcome. This rare piece of good news from Brazilian participation in international testing should not be overlooked.

We now turn to focused comparisons between Brazil and its four comparator countries to learn more about the factors that underlie the observed differences in test scores. The data in table D.7 provide a good counterpart to the spending summaries highlighted earlier (table D.4), at least for the countries that appear in both tables (Brazil, Korea, and Mexico). Compared with Korea especially, but also to some extent with the other three countries, Brazil has significant resource deficiencies. First, Brazilian student-teacher ratios are much higher than the other countries. This variable is not the same as class size, although it is related,[1] and given the centrality of teacher salaries in education expenditures (see table D.4) this provides a good indication of why per student expenditures are so much lower in Brazil.

The results in table D.7 also show a dramatic difference in learning opportunities for Brazilian students compared with their counterparts in Korea and Thailand and, to a lesser extent, Mexico and Uruguay. For example, Korean students receive upward of 50 percent more class time overall per week than students in Brazil. This difference is quite large as is, but its effect may be much larger still when one considers possible differences in the efficiency of in-class activities. There are also some additional resource gaps between Brazil and other countries.

Other comparisons using the PISA data (not presented) highlight some important environmental differences. For example, Brazilian classrooms are less ordered than in other countries. The Brazilian students report the most frequent occurrences of "students don't listen," "noise/disorder," "teacher has to wait for quiet," "students don't work well," and "work begins long after lesson begins."

Table D.8 also presents evidence about school autonomy and school governance regimes. Overall autonomy is lowest in Brazilian secondary schools, with the exception of Uruguay, where a very centralized system is in place. It is frequently argued that schools with more autonomy and control over resources are better managed. Extensive research touches on this area, and the analysis, for example, by DiGropello (2006) of secondary schooling in Latin America and Asia offers a detailed analysis using the PISA data. The variables in the bottom half of table D.8 also address these issues to some extent. One finding is that students in the other four countries are more frequently absent, according to school directors, than are students in Korea. But Brazilian schools also report more environmental problems than the other countries. These include disruptions, lack of respect, and bullying. Brazilian teachers, according to school directors, are also more likely to be absent and resistant to change than Korean and Thai teachers. These indicators also are commonly linked with autonomy and the ability of schools to control personnel.

The tables and figures in this section provide a comparative framework for considering school quality in Brazil. Several findings clearly stand out. First, when controlling for student and family background, Brazilian school

Table D.8. Teaching and Learning Environment, PISA 2003

| | Country | | | | |
Variable	Brazil	Korea, Rep. of	Thailand	Mexico	Uruguay
Autonomy over					
Resources	2.4	2.6*	3.7*	3.9*	1.9*
Curriculum	3.3	4.0*	3.9*	3.2*	1.9*
Problems reported in school					
Students absent	2.5	1.6	2.5	2.5	2.5
Student-teacher relations	1.8	1.5	1.9	2.0	1.9
Disruptions	2.5	1.7	2.0	1.9	1.5
Teachers absent	2.0	1.5	1.9	2.1	2.8
Lack of respect	2.1	1.9	1.7	1.8	1.7
Resistance to change	1.9	1.8	1.6	2.2	2.3
Bullying	2.0	1.7	1.7	2.1	1.5

Source: PISA 2003 database.

Note: Autonomy over resources is measured on a scale of 0–6. Autonomy over curriculum is measured on a scale of 0–4. Problems reported in school are measured on a scale of 1–4, as reported by school director. For all scales, low numbers = low, high numbers = high.

*Country mean is significantly different from the rest of the sample at the 0.05 level (p<=0.05, two-tail test).

efficiency is clearly inferior in mathematics and, to a lesser extent, science. This trend does not turn up for language, which is an exception noted above that has not received much attention elsewhere. Second, there are clear resource differences between Brazilian schools and others, especially in class time and teacher qualifications. Finally, the learning environment in Brazilian classrooms is not conducive to maximum achievement and is characterized by problems between students and inefficient time use.

How Students Are Taught. We now turn to a very different source of comparative information on Brazilian school performance: the study by Carnoy, Gove, and Marshall (2007) comparing Brazilian, Cuban, and Chilean primary schools. Chile is a salient comparator because it is a natural competitor for Brazil, and Cuba is useful given the high scores Cuban students obtain on standardized tests (see Carnoy and Marshall 2005). The qualitative findings of this study are based on small samples (roughly 12–15 classrooms per country), so caution is urged about generalizing to the entire country. However, the results are largely consistent with comparisons from PISA and provide greater contextual information grounded in actual classroom observations and analyses of the delivered curriculum. This is especially useful because we are trying to identify how schools are teaching the skills that have been identified as critical in the knowledge economy: communication, participation, questioning, critical thinking, and higher-order processing of information.

The data come from taped grade-three mathematics classes in each country. One instrument measured the time segments in each class while another focused on more qualitative processes within each class session. Finally, the tapes were reviewed to analyze the content of the lesson. Among the differences encountered were the following:

- Brazilian students spend much more time copying instructions and lessons from the chalkboard than do students in the other countries. This contributes to a less efficient lesson and exacerbates inequality in lesson results because the slowest students sometimes fail to complete the written activities before the lesson proper begins. In Chile, this problem is minimized by the use of photocopied problem sheets, an indicator of a resource advantage. In Cuba, the teachers frequently have parents write out student activities before class.

- Brazilian teachers rely more heavily on recitations by the whole class than on individual questions and answers. This is another potential source of unequal learning in the classroom because "class choir" activities result in a less rigorous monitoring of student progress.

- Brazilian classrooms are frequently organized in groups, but in practice, the work done is individual rather than group-oriented. This is a common finding in Latin American classrooms, where teachers are frequently disposed to using child-friendly techniques but do not fully implement the activities as intended.

- Brazilian students are noticeably (and significantly) less engaged during the lessons. This means there are more instances of talking, playing around, or generally not paying attention while the teacher is speaking. This observation-based conclusion is consistent with PISA data about children's and directors' perceptions of schools.

- In Brazilian classrooms, teachers infrequently check every student's work, and usually only check some of any student's work. This is very different from Chile and especially Cuba, where students are more likely to be asked to demonstrate competence on the lesson before moving on.

- Brazilian teachers also make much less use of direct questions to students. In some classes no questions were asked of students, and when questions were used, they tended to be rudimentary rather than probing. In no class was the teacher observed asking questions that required a conceptual or analytical response.

(The curriculum content analysis also identified some very clear differences between Brazilian classrooms and those in Chile and Cuba, which will be discussed further in the next section.)

Caveats about the sample size aside, the results from these qualitative observations of Brazilian classrooms speak volumes about the current quality of Brazil's primary schools and help fill in more gaps about why results on SAEB and PISA are so low. In sum, the classes lack the dynamic teaching required for engaging the interest of poor children and preparing them with the basic skills they need to continue learning.

What is most disturbing about this qualitative evidence is the obvious generation of inequality within the classroom. A lot of research focuses on equity issues across schools, states, or regions. But the Carnoy, Gove, and Marshall study clearly demonstrates the challenges of preparing all children within a classroom with the basic skills they need to advance. When children obviously have not completed copying the instructions by the time the lesson ends, and teachers are averse to checking every student's work or asking individual questions, it is hard not to assume that the teacher knows all too well that not every student is progressing. Even when some of the students in the classroom do move forward, the overall cognitive skills they master are undemanding and the lessons they process seem very basic. The rest not only fall farther behind their successful classmates but also watch students in other school systems race far ahead.

What Students Are Taught: The Role of Curriculum. A natural place to begin discussion of lesson materials is with the National Curriculum Parameters (PNC) defined in 1997. The PNC represent official goals or guidelines, but they are not mandated curriculum. A mandated curriculum can only be enforced through strong accountability and measurement mechanisms, which are not currently in place in Brazil (as the next section shows). The curriculum standards are not even envisioned as a complete rendering of curriculum coverage because they leave approximately 25 percent of the lesson content to be defined by the schools (that is, free).

How well is the official curriculum being implemented? The answer has obvious implications for overall efficiency and quality, as well as for equity. Given the low scores on SAEB, many students in Brazil clearly are not mastering the officially sanctioned curriculum. On its face, this would appear to be more attributable to school quality than to the curriculum per se. But if the official curriculum is spread too thinly between numerous elements, or if the introduction of different cognitive skills is poorly formulated, then low achievement has a curricular component as well.

Are schools strictly adhering to the official curriculum or are additional elements being introduced that water down the main subjects? Anecdotal evidence suggests that schools are incorporating sex education, drug prevention, and other topics into primary learning activities. These life skills are relevant to young people in Brazil, but time devoted to them may reduce time spent learning the skills needed to move onward in school or on a job.

This raises the question of curricular relevance—whether schools are targeting the kinds of skills that students will need to compete in an increasingly globalized and technically demanding labor market. In their review of labor market skills in the United States, Levy and Murnane (2004) highlight the dramatic changes in skill requirements that are occurring in the workplace. Compared with 1960, priority skills are increasingly related to specialized thinking and complex communication. So-called routine manual and cognitive skills are in less demand.

Which kinds of skills are being created by Brazilian schools? Answering this question is not easy, and sweeping statements about curriculum in Brazil are to be avoided. This is especially true given that the PNCs were introduced only in the past decade, and more time is necessary to evaluate the impact of these goals on the system. Nevertheless, the SAEB results show clear gaps between the intended and implemented curriculums. Using an international standard, the PISA confirms the breakdown, putting the negative consequences for competitiveness and future economic growth into starker perspective.

The qualitative analysis of classroom performance undertaken by Carnoy, Gove, and Marshall (2007) provides another comparative snapshot of curriculum development in Brazil based on a small cross-section of grade-three classrooms. Their analysis includes comparisons of curriculum content and goals in the Brazilian, Chilean, and Cuban mathematics lessons being observed. The results are troubling and, at the very least, provide more specific contextual detail to the school quality deficiencies identified earlier in this appendix.

The analysis of curriculum in this three-country study was conducted along four dimensions: mathematical proficiency of lesson, level of cognitive demand, format or goal of lesson, and level of support. We will focus on results for the first two areas. In the case of Brazil, all of the observed lessons (except one) possessed the basic component of conceptual understanding, or a minimum level of mathematical proficiency. The exception was a class that relied solely on rote memorization, making it impossible to rate in terms of proficiency. The Brazilian average on this construct was significantly lower than for Chile and (especially) Cuba. As the authors noted:

> The gap [in mathematics proficiency] between Cuban classroom lessons and those of Chilean and Brazilian classrooms stemmed from the use of the proficiency strands of strategic competence and adaptive reasoning. That is, Cuban teachers engage in continual dialogue with the students, asking them both how and why a given problem should be answered.

For cognitive demand, the measure is derived from work by Stein et al. (2000) in classrooms in the United States and is divided into four categories: memorization tasks and procedures without connections (both classified as lower-level demands) and procedures with connections and "doing mathematics" tasks (higher-level demands). Brazilian classrooms scored significantly lower on this construct as well:

> [In Brazil]… the lessons were focused on producing correct answers rather than developing understanding. Interestingly, when considering urban-only classrooms, Brazil's score actually decreased as the rural teachers scored higher than their urban counterparts on the cognitive demand score. This may have been due to the presence of a new curriculum and extensive training in two of the rural schools which are part of the Escola Ativa program… For the most part, Brazilian lessons consisted of a teacher writing on the board, students copying,

and little interaction. In most cases, almost no effort to link concepts to the procedure was made. Explanations, when they were made, focused solely on describing the procedure that was used.

These descriptions of content proficiency and cognitive learning in Brazilian classrooms are consistent in many ways with the results of Brazilian students on standardized tests. Evidence shows that classroom lessons are focusing almost entirely on very basic elements that do not help students develop the kinds of skills they need to be active learners and apply acquired knowledge in real-life situations.

The Accountability System

Two general explanations help explain why test scores are low and teaching and learning environments are deficient in Brazil. The first can be called the "low existing capacity" explanation, and exhibit A in this line of reasoning is the low per pupil spending. Simply stated, governments get the education systems they pay for, and in the Brazilian case, a low-quality system is to be expected.

A counterpart thesis can be termed the "low capacity maximization" explanation. In this scenario, teachers and schools are not necessarily underfunded, they are just not using their existing capacity to obtain the best possible results from available resources. Low teacher attendance, limited use of homework, frequent use of copying from the textbook—each may occur when educational agents are not properly motivated or held accountable for their actions.

These explanations are not mutually exclusive, and support for each is common in the developing world. A convincing case has already been made that Brazil lags behind others in spending, so existing capacity levels are likely to be comparatively low as well. Nevertheless, some countries (namely Korea) have also been shown to outperform others despite spending much less money. Many possible explanations for this result exist, but this section will highlight the important role that accountability systems play in determining educational performance.

Institutionally, the present accountability system in Brazil is a product of several fairly recent initiatives. Central to this schematic are standards, resources, and results. Standards describe the goals—or requirements—of the system. The previously cited curriculum standards (PNC) are critical here because they lay out in detail what Brazilian students are expected to learn by grade and subject. These goals are not accompanied by equally specific methods, however, and schools are intentionally left some room for flexibility.

We have also sketched how resources are distributed in Brazil via a complicated system with three levels (federal, state, and local). Through programs like FUNDEF, the government has worked to ensure minimum funding. Schools also receive direct aid through specific interventions like PDDE and PDE. Finally, specific support programs are in place to provide items such as textbooks and school meals and for school improvement through federal efforts such as Fundescola.

The element that ties together standards and resources can be termed results (or performance). For example, the entire system is evaluated every two years through the SAEB national standardized testing system, which uses tests closely aligned with the curricular goals laid out by the PNC. In 2005 the SAEB reached all schools rather than a sample, as in typical studies. Schools also report enrollment, repetition, and dropout rates through the school census. Each constitutes a potential metric for measuring school performance against standards or goals. Given the FUNDEF scheme, enrollment defines financing at the state and municipal levels. Performance could be used—in theory—as a means of holding schools accountable for service delivery, perhaps through the use of financial incentives for high results.

As in many countries—developed and developing alike—the Brazilian system includes individual elements keyed to accountability, but lack of coordination among the elements makes it difficult if not impossible to hold schools truly accountable. For example, the most powerful performance measure (student achievement) is collected in all schools very infrequently. Meanwhile the measures of school performance that do exist are not incorporated into funding decisions, and there is minimal formal evaluation of work by teachers and school directors. Instead, school funds are distributed mainly by a fixed funding formula based on enrollment. Teacher salaries are determined by level of education, training, and seniority and not by comparable measures of performance based on student outcomes.

Holding schools and teachers individually accountable for performance is very difficult, mainly because it requires valid measures and a credible system for evaluating the work of school personnel. The evidence from other countries, namely Chile and the United States, is mixed when it comes to measuring the impact of these kinds of policies. It is possible that low performance in Brazil is a function of low capacity rather than low maximization of capacity. This argues for spending more money or spending funds more wisely, and underlines the continued need for systemic diagnoses through activities like SAEB. Reaping additional gains by harnessing existing capacity more effectively can build on the elements of an effective accountability system already in place (SAEB, school census, local funding sources).

APPENDIX E

The Tertiary Education System and Advanced Out-of-School Training

Introduction to the Tertiary System

In 2005, the most recent year for which official statistics are available, the Brazilian tertiary education system comprised more than 2,100 institutions enrolling nearly 4.5 million students (table E.1). Gross enrollment accounted for more than a quarter of young people in this age group.

Size alone does not determine the impact of a higher education system. To assess how well the system contributes to innovation-driven economic growth, three other aspects need to be taken into account—first, access and equity (has the system expanded so that all social groups have equal opportunities to participate); second, quality and relevance (are tertiary institutions producing the kind of graduates and research outputs that the new knowledge economy requires); third, governance, financing, and management (is the governance structure appropriate to facilitate transformation of the system; is Brazil investing sufficiently at the tertiary level; are resources being allocated and utilized effectively).

Access and Equity

Coverage of and Accessibility to Tertiary Education

Brazil's tertiary education system is among the largest in the world and, paradoxically, among the least developed in Latin America. In fact, a quarter of the relevant age group in Brazil attending a tertiary institution in 2004 represents the next-to-lowest enrollment rate (followed only by Mexico) among

Table E.1. Brazil's Tertiary Education System, 2005

	Public	Private	Total
Institutions	231	1,934	2,165
Students	1.2 million	3.3 million	4.5 million

Source: Ministry of Education Web portal (March 2007).

the more developed Latin American countries and considerably below the regional average of 30.3 percent (table E.2).

Brazil's relatively low tertiary coverage is also apparent when comparisons are made beyond Latin America. For example, not too long ago China ranked far behind Brazil. Yet China has been catching up rapidly, and its coverage rate is likely to surpass Brazil's within two to three years.

Table E.3 compares the share of the labor force with tertiary education in Brazil, Chile, the Republic of Korea, Mexico, and the OECD average. With only 12 percent of 25-to-34-year-olds with tertiary education, Brazil is clearly at a disadvantage compared with its economic competitors.

Two factors stand out in explaining Brazil's low coverage. First, secondary education has grown relatively slowly. Second, the government has maintained the public tertiary education subsector at a constant size, allowing private institutions to absorb the bulk of expansion. Between 1996 and 2004, the number of public institutions grew by only 5 percent (from 211 to 224), while the number of private institutions more than doubled (from 711 to 1,789). Half the private tertiary institutions in operation today were established after 1998; and indeed, Brazil has the highest proportion of students (73 percent, see table E.1) attending private institutions in Latin America.

Equity: Who Gets In?

Not only is coverage low in Brazilian tertiary education, its equity remains a serious concern. Access to tertiary education is heavily skewed against

Table E.2. Tertiary Education Coverage in Latin America, 1980–2004
percent

Countries	1980	1990	2004	Increase 1980–2004
Argentina	21.8	38.5	63.9	292
Brazil	*11.2*	*11.3*	*25.1*	*224*
Chile	12.3	21.3	46.9	381
Colombia	8.6	13.4	27.1	315
Costa Rica	21.0	26.4	43.7	208
Cuba	17.3	20.9	41.7	241
Dominican Republic	—	20.4	36.9	—
Mexico	14.3	15.2	24.6	172
Peru	17.4	31.1	33.9	195
Uruguay	16.7	30.7	42.2	253
Venezuela, R. B. de	20.6	29.2	44.6	217
Latin America	—	15.6	30.3	—

Sources: EdStats, The World Bank, last data update June 2006, retrieved September 5, 2006; IESALC 2006.
Note: — = not available.

Table E.3. Share of Labor Force with Tertiary Education, 2004

percent

Country	25–64 age group	25–34 age group
Brazil	8	12
Chile	13	18
Korea	30	49
Mexico	16	19
OECD average	25	31

Sources: OECD. "Education at a Glance 2006." Tables, Indicator A1, accessed on November 2, 2006. www.oecd.org/edu/eag2006.

Figure E.1. Distribution of Students by Income Group

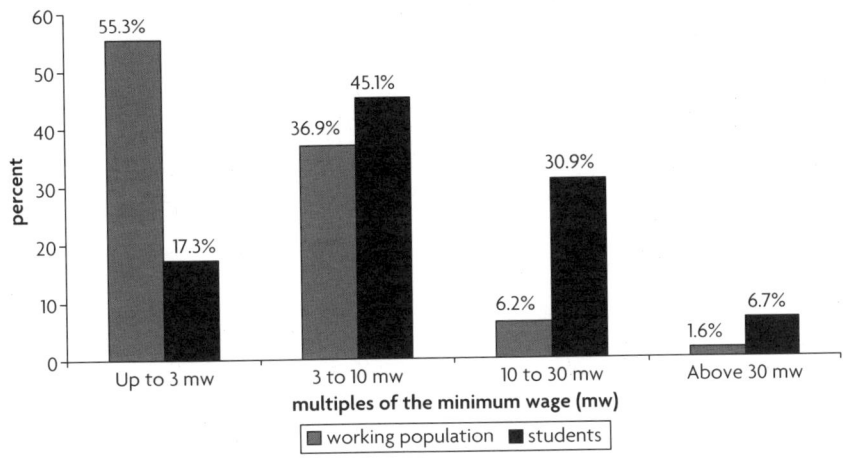

Source: JBIC 2005, with MEC/INEP data.

poorer students. Only about 5 percent of students come from the two lowest economic quintiles (2004 Household Survey, IBGE *Pesquisa Nacional por Amostra de Domicilios*). To illustrate inequality, figure E.1 compares the proportion of workers by income groups, measured as multiples of the minimum wage and the proportion of students from these same groups.

The State University of Campinas (UNICAMP) in the state of São Paulo is widely considered to be among the best universities in Brazil. Table E.4 shows the socioeconomic characteristics of UNICAMP compared with the overall state and national populations, vividly illustrating the extent of social bias in an elite Brazilian university.

Several factors have contributed to the social biases of tertiary access. First, the government-controlled system at the federal and state levels limits the number of student slots in public universities. Second, the low quality of public schools does not adequately prepare secondary students for entry

Table E.4. Socioeconomic Characteristics of Undergraduate Students at UNICAMP Compared with the State and Nation
percent of tertiary-age students

	Low-income family	Attended public high school	Father has tertiary degree	Mother has tertiary degree
UNICAMP	10	27	53	41
São Paulo	57	84	11	12
Brazil	69	83	8	9

Source: Pedrosa 2006.

and success at prestigious public universities. Third, admissions are skewed toward student applicants from private high schools—for example, two-thirds of new entrants at UNICAMP are from private high schools, compared with only 6 percent of all students from São Paulo state. Fourth, financial aid is insufficient for academically deserving students from low-income families. Brazil is a classic case of a highly regressive system. The most qualified students—that is, the children of middle- and high-income families, who usually graduate from exclusive private secondary schools—fill the ranks of the free top public universities. Students less academically qualified, from families much less well off, are limited to fee-charging, private, tertiary education institutions.

Equity Improvement Programs in Tertiary Education

ProUni. In 2004, the Ministry of Education launched ProUni (*Programa Universidade para Todos*), a program to place academically qualified low-income students into private universities. To qualify, a student must have a passing grade in the voluntary national end-of-secondary-school examination (ENEM) and demonstrate that he/she comes from a low-income family. Full-time students who receive the full scholarship are also eligible for a monthly maintenance grant of R$300. The government's program target is to finance up to 400,000 students by 2008, which would be equivalent to 35 percent of the current public university enrollment.

In practice, there is no actual transfer of resources from the Ministry of Education to participating universities. Instead, the universities receive a tax exemption up front during the first year of participation in the program. The tax exemption continues every year as long as scholarships are maintained for students that entered in previous enrollment rounds, and provided that scholarships for new students who qualify for ProUni constitute one of every 10.7 new enrollments (one of every nine in the case of a nonprofit university).

Even though ProUni is only in its third year of operation, the program has produced tangible benefits. About 120,000 students benefited during the first year (out of 340,000 candidates), and another batch of 91,000 was enrolled during the first part of the 2005/06 academic year (out of 800,000 candidates). The drastic increase in the number of candidates has allowed the Ministry to raise the cut-off point for eligibility based on student ENEM results from 56 in 2004/05, to 62 in 2005/06.

Notwithstanding its innovative character and ingenuous financial design, ProUni raises a number of questions requiring further investigation before the program's effectiveness and impact can be fully assessed: (a) Is there proper targeting? (b) What is the quality of the participating private universities? (c) What is the actual opportunity cost of the tax exemption? (d) Is ProUni reinforcing the existing unequal pattern of tertiary education development?

Affirmative Action Programs. The government of Brazil increasingly has been concerned with racial inequities reflected in the education system. In a country where 6.2 percent of the people consider themselves black, only 2 percent of the university student population is black. To address this issue, the government submitted legislation to Congress that would oblige the federal universities to reserve at least half of new places for students originating from public schools to be divided among black, mixed race, and indigenous students.

Notwithstanding the good intentions behind these quota programs, the government of Brazil needs to carefully review international experience with affirmative action programs, which has highlighted the following challenges: (Sowell 2004):

- How to limit preferences and quotas in time and scope

- How to ensure that the actual beneficiaries are those targeted by the preference program

- How to avoid polarization leading to intergroup resentment and conflict

- How to avert overall efficiency losses.

Student Loan Programs. Brazil has had a national student loan program since 1976. The program has been managed by the Federal Savings Bank (*Caixa Economica Federal*, CEF) and has evolved through different structures over time. It ran into serious difficulties in the early 1990s because of excessive defaults (up to 70 percent of loan recipients) from high inflation rates and ineffective claims procedures against defaulters.

The student loan program was cancelled in 1994, and a new scheme was set up in 1997 as the *Fundo de Financiamento ao Estudante do Ensino Superior* (FIES). FIES loans finance 50 percent of tuition fees (reduced from 70 percent in earlier years) at a fixed annual interest rate of 6.5 percent or 3.5 percent, depending on the program of study (down from 9 percent since 2006). Participating students, who need to have two guarantors (except in Alagoas state), can enroll only in institutions accredited by the Ministry of Education and must maintain good grades (a 75 percent average) to continue benefiting from the loan. By 2006, about 390,000 students had received a FIES loan.

The administrative setup of FIES seems reasonably lean. The program is supervised by a small group within the Ministry of Education and administered by CEF on behalf of the ministry. Because ProUni handles financial aid for the poorest students, it can be assumed that FIES is reaching the majority of non-ProUni students who need financial assistance. Verification, however, requires an appropriate survey.

Notwithstanding the positive features of FIES in terms of management and coverage, several adjustments could be considered. The first concern involves eligibility. Because there is no family income ceiling to qualify, there is a risk that students from wealthy families could take advantage of the 6.5 percent concessional interest rate to support expenditures not directly linked to their studies (because money is fungible). Second, the loan covers only half the tuition fees, and low-income students may not be able to fund the other half themselves. They may also be unable to cover living expenditures if they are not working. Third, the repayment schedule provides for equal monthly installments. This has the major drawback of constituting a relatively high proportion of income at the beginning of a graduate's professional career and a relatively smaller burden as postgraduate income increases with time. Finally, the financial sustainability of the student loan system needs to be investigated. So far, payment compliance has been satisfactory, with only 11 percent of graduates defaulting on their loans. But it is still important to monitor the accumulated costs of the program—namely the 2 percent administrative fee to Caixa, the cost of nonpayments, and the interest rate subsidy—to ensure that the FIES program does not decapitalize.

Out-of-School Advanced Skills Training

Brazil has developed an extensive out-of-school skills training system in the years since the mid-twentieth century industrial boom. Today the system comprises a group of institutes, collectively known as the S-system, offering advanced skills training and other services for workers. The earliest S-system institutes focused on training and services for industrial workers, but over the years, new institutes emerged to offer skills training in agriculture, commerce, and small business. Despite the strengths of this approach, Brazil needs to establish stronger links between the S-system and the tertiary education system. The disconnect is hampering cross-fertilization between academia and business and slowing Brazil's potential for disseminating and commercializing new knowledge, significant shortcomings from an innovation perspective. At the same time, few other advanced skills training options have emerged.

Vocational Education: The Brazilian S-System

Brazil's S-System is the largest consolidated professional training system in Latin America. The system initially emerged from the industrial sector in the 1940s, when the National Confederation of Industry (CNI) joined with state-level industry federations to lobby for the establishment of SENAI, the National Service for Industrial Apprenticeship, and SESI, the Social Service for Industry. The objective of SENAI was to train and qualify personnel for industrial sector jobs; SESI offered social services to improve the quality of life for industrial workers. The S-system was subsequently expanded with the establishment of parallel institutions serving the commercial, transportation, agriculture, and worker cooperative sectors. Today the system is not a single entity but an assemblage of nine separate, loosely related national institutes organized by sector. With a presence in all 26 states of the nation and in the Federal District, the system operates in more than 3,000 municipalities through nearly 5,000 units and attendance points. Participants in its training and social service activities can be grouped into three broad categories—a small group of youth up to the age of 18 who are concentrated primarily in apprenticeship and training; a large group of youth between the ages of 18 and 30 who are unemployed, working in the informal market, or seeking to improve their technical skills; and an equally large contingent of workers between the ages of 20 and 40 who are directly sponsored by their employers to receive training. The nine institutes composing the S-System include (a) the National Service for Industrial Apprenticeship (SENAI), (b) the National Service for Commercial Apprenticeship (SENAC), (c) the Social Service for Commerce (SESC), (d) the Social Service for Industry (SESI), (e) the Brazilian Service for Assistance to Small Business (SEBRAE), (f) the National Service for Agriculture Apprenticeship (SENAR), (g) the Social Service for Transport Industries (SEST), (h) the National Transport Apprenticeship Service (SENAT), and (i) the National Apprenticeship Service in Cooperative Activities (SESCOOP). (See below for further information on each of the national institutes.)

Taken together, the S-System offers an estimated 2,300 courses per year, with an annual enrollment of roughly 15.4 million students. In 2006, the combined budget was projected at more than R$13 billion. A compulsory 2.5 percent payroll tax on private companies supplies 85 percent of the budget; the rest comes through contracts with the public sector, informal relationships with companies, unions, mayors, and communities, as well as through out-of-pocket expenses paid by participants. The components of the S-System are briefly summarized below.

National Service for Industrial Apprenticeship (SENAI). SENAI was set up on January 22, 1942, by Decree-Law No. 4,048 of President Getúlio Vargas to train manpower for the basic industries that were about to be launched. Without such vocational training, industrial development in Brazil would be stillborn. Over the next 20 years (from the 1940s to the end of the 1950s) SENAI became a reference point for innovation and quality in vocational training, serving as a model for similar institutions in Argentina, Chile, Peru, and the República Bolivariana de Venezuela.

In the 1960s, SENAI invested in systematic training courses, increased on-the-job training, and sought partnerships with the Ministries of Education and Labor and with the National Homestead Bank. During the economic crisis of the 1980s, SENAI recognized the major shift under way in the economy and decided to invest in technology and in developing its specialist staff. It increased the assistance it offered to companies, acquired the latest technology, and set up teaching centers for research and technological development. With the technical and financial support of institutions in Canada, France, Germany, Italy, Japan, Switzerland, and the United States, SENAI entered the 1990s ready to advise Brazilian industry in production technology, product design, and business management.

From an average of 15,000 students in the early years, enrollments have grown to about 2 million annually, totaling approximately 39 million enrollments since 1942. The first handful of schools became a network of 744 operational units distributed nationwide, which today offer more than 1,800 courses and more than 80,000 technical and technological advisory services per year to companies. Currently, SENAI has 27 regional departments, all linked to a national department. It offers courses through the following conduits:

- Vocational education centers—236 vocational education units develop courses and programs in different types of education for young people and adults, as well as attending to the production sector.

- Technology centers—43 vocational education units transfer technology through training, provision of technical services, and the diffusion of information about technology.

- Mobile units—316 vocational education units provide SENAI services in regions far from Brazil's centers of production. In addition to a river unit, SENAI has a fleet of trailers and vehicles that act as real traveling schools.

- Mobile Activities Program (PAM)—310 PAM teaching teams operate as portable workshops. The PAM kits were specially designed to reach the remotest parts of the country with programs in 27 vocational areas.

The National Service for Commercial Apprenticeship (SENAC). *Serviço Nacional de Aprendizagem Comercial* is a vocational educational institution working in the commerce and services sector. It was created by the National Confederation of Business (CNC) on January 10, 1946, through Decree-Law Nos. 8,621 and 8,622.

During its 58 years of operation SENAC has trained more than 40 million people in the commercial and service trades, helping to raise respect for workers through vocational training in 12 areas: the arts, commerce, communication, conservation and curatorship, design, management, personal image, computing, leisure and social development, environment, health, and tourism and hospitality.

SENAC currently operates in nearly 2,000 municipalities, offering 1.8 million trainees access to a spectrum of educational opportunities—(a) classroom-based courses; (b) distance learning, which includes correspondence courses as well as television- and radio-taught courses; (c) part-time courses involving two different but complementary phases (alternating direct contact between teacher and student with periods of independent study guided from a distance by the teacher); and (d) the SENAC Móvel (Mobile SENAC) Program, which sends mobile educational units all over Brazil, carrying educational infrastructure to the remotest regions. Mobile units stay in each municipality from six months to a year, and their stays are arranged through partnerships with local councils, state governments, or bodies representing community interests.

The Social Service for Commerce (SESC). The Social Service for Commerce (*Serviço Social do Comércio*, SESC) was created by Decree-Law No. 9,853 in 1946. It is supported by employers in retail trade and services to promote the social well-being of its workforce through enhanced education, health, leisure, culture, and social assistance. The organization also assists those living on the edges of small, medium, and large towns to form partnerships with public service providers, private firms, trade unions, and residents' associations.

Today, SESC serves about 3.6 million people who are mostly workers in the goods and services sector and their families and dependents. It is found in all state capitals in Brazil and in small and medium-size towns.

Education is SESC's historic mission and is identified as the essential path to lead workers and their families to a better quality of life. Various activities are designed to engage children, teenagers, and adults in active citizenship. Social and educational activities include nurseries, early childhood education, primary education, adult education, preparation for university entrance examinations, preventative and supportive medicine, dentistry, nutrition, cinema, theater, the plastic arts, dance, crafts, libraries, sport, community action, and targeted assistance.

The Social Service for Industry (SESI). *Serviço Social da Indústria* was created in 1946 to improve the quality of life of industrial workers and their families. SESI sponsors activities in basic and complementary education, medical and dental care, leisure, sports and culture, and other efforts of social benefit.

In addition to the services provided through its 324 activity centers and 891 operational units and 748 mobile units, its Regional Department develops activities inside industrial firms tailored to the employer's needs and expectations. Various projects benefit the community through partnerships and agreements with national and international governments and private institutions.

SESI's 1,963 units are distributed across 2,006 municipalities in 27 states, including physical infrastructure that includes 11,701 classrooms, 1,229 dental offices, 150 laboratories, 127 workers' clubs, 198 fitness centers, 64 stadiums, 184 auditoriums/cinemas/theaters, 8 vacation camps, 527 swimming pools, 623 sports venues, 312 football fields, and 80 industrial kitchens.

The Brazilian Service for Assistance to Small Business (SEBRAE). SEBRAE, originally CEBRAE, was created in 1972 to improve the business climate for small businesses in Brazil. As of 2003, its priorities included (a) tax reductions; (b) decreased bureaucracy; and (c) greater access to credit, technology, and knowledge. Currently, SEBRAE is present in all 26 states and the Federal

District, with more than 600 points of attendance distributed across the country from the northernmost to the southernmost boundaries. SEBRAE offers training, facilitates access to financial services, promotes cooperation between businesses, organizes work fairs, and serves as an information clearinghouse for small businesses.

The National Service for Agriculture Apprenticeship (SENAR). SENAR was created by Law No. 8,315 in 1991. Linked to the Brazilian Confederation of Fish and Agriculture (CNA), SENAR is charged with nationally organizing, administering, and implementing Rural Professional Training (FPR) and Social Promotion (PS) programs for youth and adults in farming areas. Program activities focus on strengthening the self-esteem and technical skills of rural laborers through multidisciplinary teams that design and teach relevant courses. At the end of each course, participants receive documentation to certify their satisfactory participation and learning.

The Social Service for Transport (SEST/SENAT). *Serviço Social de Transporte* and the National Transport Apprenticeship Service (*Serviço Nacional de Aprendizagem do Transporte*) were created in 1995 to "develop and disseminate the culture of transport, improving its workers' quality of life and job performance, as well as training new workers to provide efficient and quality services beneficial to society."

The two organizations have 96 units distributed through all 27 states. SEST offers products and services such as (a) basic dental treatment; (b) medical treatment (gynecology, pediatrics, ophthalmology, and general clinical care); and (c) leisure activities, sports, and culture to meet the needs of transport workers, their families, and the community. SEST plans and funds numerous projects for social inclusion at the municipal, state, and federal levels, including campaigns serving the elderly, women, and expectant mothers. SENAT offers supplementary education at the primary and secondary levels and training and certification programs for transport workers.

The National Apprenticeship Service in Cooperative Activities (SESCOOP). *Serviço Nacional de Aprendizagem do Cooperativismo* originated as RECOOP, the Recovery Program for Cooperative Activity in Agriculture (*Programa de Recuperação do Cooperativismo Agropecuário*), which was founded to organize, administer, and carry out vocational training, development, and social advancement of cooperative members throughout Brazil.

The National Apprentice Service in Cooperative Activities was created in 1998 along the lines of the other eight centers in the Brazilian S-System (SENAI, SESI, SENAC, SESC, SENAT, SEST, SENAR, and SEBRAE), which coalesce private sector initiatives to develop vocational training programs by productive sector. What makes SESCOOP unique is its focus on those involved in cooperatives, tailoring its techniques and goals to raising productivity and management in cooperative societies.

Lifelong Learning

Brazil does not have a lifelong learning strategy yet. As discussed above, the country has a well-developed network of vocational training institutions—the S-System—but there are very few linkages between it and the tertiary education system administered by the Ministry of Education. Even within the tertiary education system, mobility across different institutional types is limited. Few institutions have modular course organization based on academic credits that would facilitate transfer from one kind of institution to another. There is no recognition of prior or on-the-job learning. Career guidance is not well developed, and there are no special financing mechanisms for lifelong learners.

Short Courses for Adults

A key dimension of a lifelong learning system is the opportunity to enroll in short professional programs—for example, offerings modeled on those in the French institutes of technology or the coursework in North American community colleges that has highly practical content directly linked to local labor market requirements. Historically, very few institutions and programs of this sort have operated in Brazil. New legislation was passed in 1996 that opened the door to two kinds of short-duration programs—technology courses and sequential courses within existing programs. The technology courses, which usually cover two-and-a-half years, can be offered through either tertiary education institutions or specialized training centers. They lead to a degree that allows the holder to continue postgraduate training. The sequential courses, which take up to two years, are offered as part of regular four-year programs, with the student receiving a certificate of study upon completion.

The number of these short programs has grown slowly, constituting only 2 percent of overall enrollment by 2003. A 2003 survey of technology training courses (JBIC 2005) confirms that these programs do indeed perform a critical lifelong learning role and offer educational opportunities to unconventional students.

Distance Learning

The final dimension to consider in this context is the availability of distance education as a flexible modality for employed youths to study part-time. Distance education appears to be at a very early stage of development. In 2004, it enrolled a mere 1.4 percent of all students.

In sum, while important components are in place to develop a lifelong learning strategy, a major roadblock remains the lack of a framework for recognition of skills acquired through formal schooling, the S-system, or on-the-job experience. This flexible framework is essential because adults are less likely to invest in continuously upgrading their skills by any of the host of alternative means available if they cannot be certain that their learning is certified and carries cachet in the labor market. Many developed nations have created such frameworks, and other developing nations such as Chile are in the process of doing so.

Quality and Relevance of Brazilian Universities

Brazil boasts a small number of excellent universities among its 2,000 institutions of higher learning. The top-five federal and state universities account for a large proportion of scientific research carried out in the country and most of the better national graduate programs. A single university, UNICAMP, accounts for about 15 percent of all scientific output in Brazil and 10 percent of all postgraduate degrees. The 1,800 private institutions range from first-rate universities engaged in research and teaching, such as the Catholic universities of Rio de Janeiro (PUCR) and São Paulo (PUCSP), to a multitude of single-faculty institutions of variable standards. Many of the smaller public institutions also are considered to be of average quality.

Are Brazilian Universities World Class?

Notwithstanding the methodological limitations of any ranking exercise, international league tables show that the world's highest ranked universities are those that make significant contributions to the advancement of knowledge through research; teach with the most innovative curricula and pedagogical methods under the circumstances most conducive to learning; make research an integral component of undergraduate teaching; and produce graduates who stand out because of their success in intensely competitive arenas during their education and, more important, after graduation. These concrete accomplishments and the international reputation that accrues from them make a university "world class."[1]

How do Brazilian universities stack up against the world's best, and against Latin American universities in particular? Two prominent international rankings have emerged since 2003. Aside from the merits of individual rankings, both are useful in comparing the priority and support for tertiary education among countries of similar economic development, population, political stability, and other indices.

First, the United Kingdom's *Times Higher Education Supplement* (THES) ranks the top 200 universities in the world. Although no Brazilian institution was included in the 2004 THES ranking, the University of São Paolo cracked the list in 2005 by landing at 196 before slipping out in 2006. The only other Latin American institution to make the THES ranking is UNAM, the National Autonomous University of Mexico, which ranked 195th in 2004, rose to 95th in 2005, and reached 74th in 2006. By comparison, four Chinese universities rank in the top 100 (at 15, 62, 72, and 93). India's Institutes of Technology and China's Institutes of Management, which are multicampus institutions, ranked 57th and 84th, respectively, in 2006.

Second, Shanghai Jiao Tong University in China has developed its own World University Rankings into clusters using a methodology that employs

Table E.5. World University Rankings by the *Times Higher Education Supplement*, 2006

Country	Number of THES top-200 institutions	Ranking positions
Brazil	0	*none*
China	6	14, 28, 116, 165, 179, 180
India	3[a]	57, 84, 183
Korea, Rep. of	3	63, 150, 198
Russian Federation	2	93, 164
Mexico	1	74
Argentina	0	*none*

Source: Times Higher Education Supplement 2006.
a. Two of India's ranked universities (the Institutes of Technology and Institutes of Management) are multicampus institutions.

seemingly objective indicators, such as the academic and research performance of faculty, alumni, and staff. Shanghai's 2005 ranking of the top 500 universities worldwide includes seven Latin American universities, four of which are from Brazil—the University of São Paolo (101–52 cluster), the University Estadual Campinas (203–300), the Federal University of Rio de Janeiro (301–400), and the University Estadual Paulista (401–500). Mexico's UNAM is ranked lower (153–202) than Brazil's University of São Paulo. Argentina's University of Buenos Aires (UBA) is ranked in the 203–300 cluster, and the University of Chile is ranked in the 301–400 cluster. By contrast, China has eight institutions in this ranking, India has three (with its management and technology institutions including multiple campuses), and Korea has seven.

Comparing Brazil's place in such rankings provides an interesting perspective on the position of its universities in the broad context of international tertiary education. Even though Brazil is the fifth most populous nation and the eighth largest economy on the planet, unlike China and India it has no university ranked among the world's top 100 in either assessment. In the more subjective, reputation-based survey (THES), in the only year it appeared, Brazil's University of São Paolo was listed lower than it was in the Shanghai ranking. This may indicate the limited exposure of Brazilian tertiary education to a broad international audience (perhaps due to a greater language barrier or more-limited exchanges of faculty and students than is the case for comparator countries).

Higher Education and the Brazilian Economy's Need for Competitiveness

To increase economic competitiveness, Brazil must better align its tertiary education system with the job market. Too few students are acquiring relevant skills and knowledge at the undergraduate level, and there are limited opportunities to pursue cutting-edge research at the graduate level, especially in science, technology, and business programs. Those graduate programs that do pursue cutting-edge research tend to emphasize theoretical knowledge and academic publication rather than patents and knowledge with commercial potential.

Graduate Unemployment. Unemployment has been rising steadily in Brazil, from about 3 percent in 1993 to 9.3 percent in 2004. According to the 2004 Household Survey, the unemployment rate among university graduates was 16.4 percent, almost double the national average. In recent years, rising unemployment among university graduates has become a serious concern, reflecting a potential mismatch between the supply of graduates and labor market needs.

Skills Mismatch. One of the most worrisome features of the Brazilian tertiary education system is the lack of priority assigned to science and technology programs. Although numbers vary with the method of classification, it is clear from figure E.2 that science and engineering receive insufficient emphasis in Brazil.

Analysis of graduates' distribution by discipline shows that the social sciences accounted for 65 percent of all undergraduate degrees at public institutions and 75 percent of all undergraduate degrees at private institutions in 2003. By contrast, engineering, sciences, mathematics, and computing accounted for a mere 18 percent of graduates in public universities and only 11 percent in

Figure E.2. Proportion of Students Enrolled in Science and Engineering Programs in Selected Latin American Countries

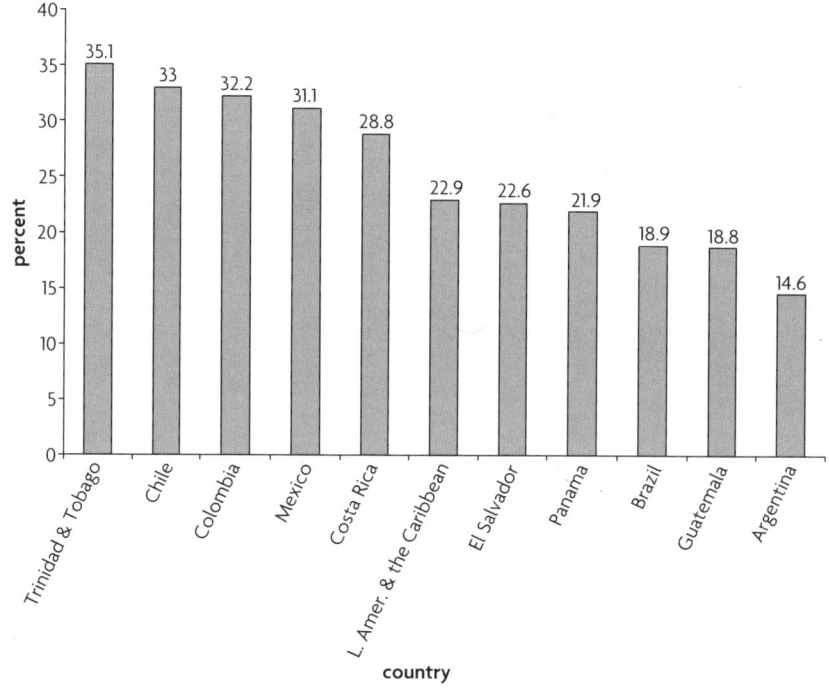

Source: IESALC 2006.

private institutions. The JIBC study (2005) attributes this pattern to a combination of supply and demand factors. Most enrollment expansion in Brazil, for instance, has occurred in private sector institutions opting to offer "soft" career paths that do not require heavy institutional investments. Industrial growth and job opportunities have not warranted such investments.

Another potential mismatch is the imbalance between degree-level and technician-level qualifications. As discussed, short professional courses account for just 2 percent of Brazilian student enrollment, far below the rates of other Latin American countries or OECD countries.

Another reason for slow growth in technical disciplines is that they are not well regarded in Brazilian society. It is more prestigious to enroll in a regular university program than in technology studies. The competition to enter technical programs, even though they are much shorter, is much less intense than for access to traditional universities. For 2003, INEP statistics indicate an average of 2.4 candidates for every university place versus 1.8 candidates for places in technology studies.

Business Education. Brazilian MBA programs also do not fare particularly well in international or Latin American regional rankings. Of the many MBA programs in Brazil, only the COPPEAD Graduate School of Business at the Federal University of Rio de Janeiro made the Financial Times world ranking of MBAs (92nd place). Three Chinese programs ranked much higher. The regional business magazine *América Economía* also listed COPPEAD as Brazil's top management program. However, it ranked COPPEAD only ninth regionwide, with programs from Argentina, Chile, Costa Rica, and Mexico finishing higher.

Graduate Programs and University Research. CAPES (*Coordenação de Aperfeiçoamento de Pessoal de Nível Superior*), which operates under the authority of the Ministry of Education, has primary responsibility for financing and evaluating postgraduate studies, disseminating scientific research, and promoting international scientific cooperation. CAPES has played a crucial role in the rapid expansion of postgraduate programs during the past decade, with the number of master's students growing from about 44,000 in 1996 to 65,000 in 2003, and the number of doctoral students rising from about 20,000 to 37,000 in the same time frame. The increases have coincided with better geographical distribution of postgraduate courses, resulting in significantly fewer regional disparities in opportunities for advanced training. Despite the recent rapid expansion, postgraduate student enrollment remains relatively low compared with other countries in Latin America. Figure E.3 shows that Brazil's 2.8 percent enrollment rate (2003) was only about half of Mexico's or Colombia's.

One positive feature is that program distribution at the masters and doctoral levels is much more balanced than for undergraduate studies. The humanities and social sciences account for about 28.5 percent of the total enrollment in masters programs and about 26.8 percent in doctoral

Figure E.3. Proportion of Graduate Students in Selected Latin American Countries, 2003

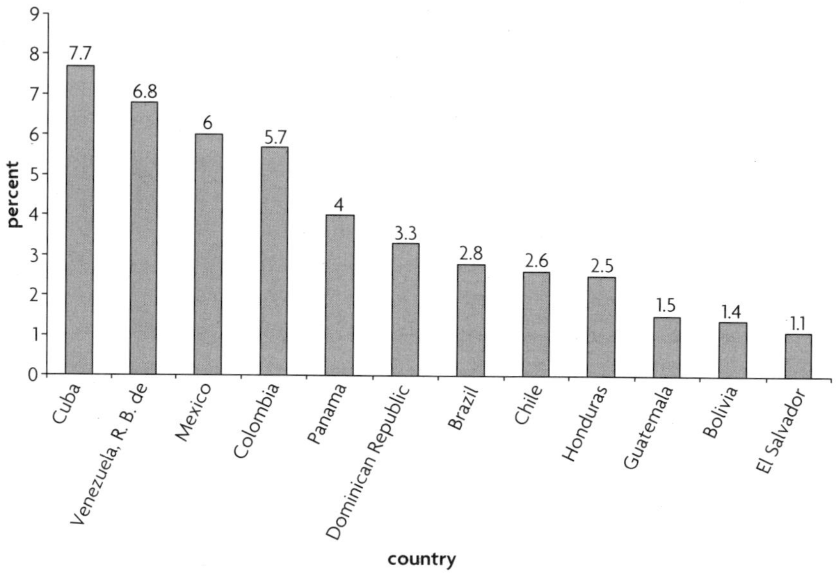

Source: IESALC-UNESCO 2006.

programs. For undergraduate studies in public universities, the corresponding proportion is 65 percent.

University Research Output

Brazil is the primary contributor of research products in Latin America. The number of patents it registered in the United States almost doubled from 63 in 1995 to 106 in 2004. The output of scientific publications kept pace, increasing from 2.2 articles per 100,000 inhabitants in 1995 to 4.1 in 2001. Brazil was outperformed, however, in the latter by Argentina (8.1 publications per 100,000 people in 2004), Chile (also 8.1), and Uruguay (4.6). Similarly, Brazil's rate of patent acquisition was far outstripped for 1995–2004 by Korea, which quadrupled its rate, and China, which posted a sixfold increase.

Contrary to the pattern in industrial countries, most researchers in Brazil are employed in the university sector (70 percent of new PhDs were hired by universities in the 1990s). Contrary to the trend in most countries, the proportion of PhDs working in firms actually has decreased in recent years.[2]

The national averages mask important disparities among institutions. In reality, research is concentrated at a very small number of universities. Three universities in São Paulo state (USP, UNICAMP, and UNESP) account for roughly half of Brazil's total scientific production. In considering research output, the Brazilian university sector basically can be divided into three groups: (1) three to five top institutions that are very productive and maintain research quality at the leading edge internationally; (2) five to ten universities that are reasonably productive in specific fields; and (3) the majority of institutions, which conduct little if any research (despite widespread aspiration to be recognized as research universities). In many cases, "research universities" operate more as university colleges or even as community colleges. To illustrate the skewed distribution, table E.6 presents the results of a recent CAPES evaluation to identify research universities with programs considered to be world class (levels six and seven on the CAPES evaluation scale).

In general, research in Brazilian universities tends to be mostly theoretical in nature, partly because CAPES evaluations emphasize the publication of articles in scientific journals. Very few institutions have managed to forge close links with industry. UNICAMP, for instance, operates a self-financed Innovation Agency that has been quite successful in creating a culture of applied research, helping researchers to register as many as 30 patents between 2004 and 2007. The Catholic University of Rio de Janeiro, which is strong in the area of computer and software engineering, has established a flourishing incubator for business applications. USP has an excellent record in biotechnology, energy, informatics, and engineering.

Table E.6. Research Universities with at Least Two Programs Highly Ranked Internationally

Acronym	Institution	Number of highly rated programs
USP	Universidade de São Paulo	55
UFRJ	Universidade Federal do Rio de Janeiro	25
UNICAMP	Universidade Estadual de Campinas	23
UFMG	Universidade Federal de Minas Gerais	14
UFRGS	Universidade Federal do Rio Grande do Sul	13
UNIFESP	Universidade Federal do São Paulo	8
UFV	Universidade Federal de Viçosa	8
PUC-RIO	Pontificia Universidade Católica do Rio de Janeiro	6
UFSC	Universidade Federal de Santa Catarina	5
UNB	Universidade de Brasília	4
UNESP	Universidade Estadual Paulista Júlio de Mesquita Filho	4
UFSCAR	Universidade Federal de São Carlos	3
UFF	Unniversidade Federal Fluminense	3
UFSM	Universidade Federal de Santa Maria	2
UFC	Universidade Federal do Ceará	2
UFBA	Universidade Federal da Bahia	2
UFPR	Universidade Federal do Paraná	2
INPE	Instituto Nacional de Pesquisas Espaciais	2

Source: CAPES.

Governance and Financing

Brazil needs greater legal and administrative flexibility in its tertiary system as well as stronger performance incentives in its federal support for public universities. A restrictive legal framework prevents public universities from making flexible and effective use of the resources they have, and federal funding provides few incentives to increase efficiency, much less to produce patents or pursue research with commercial potential. Even at the macro level, federal financing is ineffectively distributed, going mainly to public universities attended by a small percentage of the population.

Governance and Management

The Ministry of Education's Secretary for Higher Education (SESU) is the main body in charge of steering and managing tertiary education in Brazil. Its mission is to plan, coordinate, and supervise implementation of higher-education policies. Three semi-independent agencies complement the work of SESU. CAPES is responsible for the development and improvement of postgraduate training and research. INEP collects data and publishes statistics on tertiary education institutions. CNPq (the National Council for Scientific Research) coordinates and funds research activities in public and private universities.

While SESU determines policies for the entire sector, the federal government has no direct jurisdiction over state and municipal tertiary education institutions. The various higher education councils of Brazil's states make all management decisions pertaining to their institutions' budgets, personnel, salary policies, student admissions, the status of new institutions, and so forth. State and municipal institutions are required to follow federal guidelines only in curriculum because only the national government can certify diplomas.

Article 207 of Brazil's 1988 Constitution guarantees university autonomy in pedagogical, scientific, administrative, and financial matters; and the 1996 National Education Law (*Lei de Diretrizes e Bases*, LDB) provides universities with the freedom to set their own personnel policies, establish research programs, adjust their enrollments to capacity, and enter into contracts as legal entities. However, these principles of autonomy are undermined by the plethora of laws, decrees, resolutions, and regulations that organize the tertiary education sector and define how universities actually may operate. In the words of N.B.S. Ranieri (2006), a legal expert writing on the impact of the Brazilian higher-education legal framework, "In reality, the more the legislation attempts to discipline and regulate the higher-education system, the less the state is able to expand its range of action and mobilize the instruments that are at its disposition to achieve its desired objectives; and the more it legislates, the less internal consistency there is. From this perspective, it appears that the Law, as far as university autonomy is concerned, does not fulfill its function of providing incentives and stimulating socially desirable behaviors, notwithstanding the plethora of organizational norms."

Comparatively, public universities in Brazil appear to have less autonomy than counterpart institutions in, say, OECD countries. Among the more salient differences are the right to borrow from commercial banks, the ability to create positions to hire new teaching staff, the flexibility to offer competitive remuneration, and the authority to dismiss nonperforming staff members.

On the other hand, these restrictions do not extend to all public universities in Brazil. São Paulo State's universities enjoy greater flexibility—including the right to decide on the number of new positions and the right to increase the salaries of better-performing academic staff. Indeed, this flexibility goes a long way toward explaining one of the most striking features of the Brazilian tertiary education system, namely that the top two universities (USP and UNICAMP) are not federal institutions. Generally speaking, public universities in Brazil are subject to administrative rigidity that constrains the management of their resources and prevents them from operating with the flexibility that universities enjoy in other parts of the world. For example, although professors are hired through open competition, the federal and state governments control the number of positions. The salary scale is the same throughout Brazil, and promotion is based on years in service, not performance. It is difficult to recruit part-time practitioners from industry and almost impossible to cross-fertilize departments by bringing in visiting professors for a term (much less a full academic year) from public universities in other states.

In selecting a rector, however, public universities have almost full autonomy. University rectors are appointed by the president of the Republic. Three candidates, who must hold at least a master's degree, are elected by the entire university community, including students, administrative personnel, and teachers (with the latter maintaining 70 percent voting power). Rectors serve a renewable four-year term. As in other countries, the electoral dimension of the selection process introduces issues of political clientelism.

Financing

Federal support for tertiary education is unequally distributed and inefficiently used. A large pool of resources goes to educate a relatively small number of students at public universities, and funding is not linked to productivity.

Resource Mobilization. At 54 percent (in 2005), tertiary education's share of federal education spending is more than almost any other country's in the world. This unusually high proportion reflects two factors. First, the financing of primary and secondary education is shared between the federal government and the state governments. Second, the federal universities historically have been financed generously by the federal government, often without concern for efficiency in the deployment and use of resources.

All in all, the Brazilian government devotes the equivalent of 1 percent of GDP to tertiary education, a little bit less than the 1.3 percent OECD average. This level of public spending certainly seems high considering, first, the low level of enrollment in tertiary education in general and, second, that three-quarters of students attend private universities at their own expense. A major determinant of this relatively high public expenditure is that public university tuition is heavily subsidized. In accordance with the 1988 Constitution, all federal, state, and municipal public universities are free of charge. The federal tertiary institutions generate less than 3.5 percent of their total resources (Schwartzman and Castro 2005).

Resource Allocation. Until the mid-1990s, the budgets for public tertiary education institutions were utterly "de-linked" from performance. Like many countries in the developing world, Brazil had negotiated an allocation system to distribute the budget among federal and state public universities. In 1997, the federal government enacted measures to encourage tertiary education institutions to be more efficient, linking their financial resources to objective indicators such as the number of students and postgraduate activities. The impact of these measures has been mitigated by the disproportionate share of salaries and pensions in the budgets of each federal university. For example, personnel expenditures grew from 77.6 percent of the total budget transferred to the federal universities in 1995 to about 85.2 percent in 1992 (JIBC 2005).

Generally speaking, the distribution of funding of public universities takes neither institutional nor individual performance and productivity into account. Universities receive funding whether or not they perform well, produce employable graduates, or efficiently use their resources. Other than intrinsic motivation and perhaps personal allegiance to the task of nation building, individual faculty members have few incentives to improve their research and teaching. As civil servants, their positions are secure. Their success is not bound to the impact of their scholarship, research, or the competitive capacities of the students whom they train.

Resource allocation occurs in a somewhat more transparent, objective manner at the postgraduate level. The scholarships given by CAPES and the research grants available from the Ministry of Science and Technology (through CNPq and FINEP) are allocated competitively, based on the quality of programs and research proposals.

As a demand-side mechanism, ProUni is the other atypical mechanism for resource transfer within the tertiary education system (even though, as noted earlier, no additional money goes directly to universities because the purchase of "seats" for low-income students is financed through tax exemptions). It is worth noting that few countries in the world allocate public resources to universities through a demand-side mechanism as transparent and objective as ProUni. Kazakhstan and Georgia in Central Asia and the state of Colorado in the United States provide vouchers to university students—the only other examples of demand-side schemes to finance recurrent expenditures in tertiary education.

Resource Utilization. Brazil's tertiary education system has long had a reputation for high unit costs, especially in the federal universities. A recent UNESCO study shows the extent to which Brazil is an outlier in Latin

America (figure E.4). The data indicate that Brazil's unit costs are at least twice as high as those of Colombia and Cuba (the most expensive systems in the region) and three times as expensive as those of Argentina, Mexico, or Uruguay.

Two factors primarily explain the extraordinarily high costs—first, an exceptionally low student-teacher ratio of 11.4 to 1 in 2004 (despite significant improvements since the mid-1990s); and high personnel costs. Not only is the number of teachers excessive relative to the number of students, but public universities are also financially responsible for pensions of their retired professors. Brazil has a generous pension system. Professors are allowed to retire at 100 percent of their salary after 25 years of service. As a result, the proportion of the personnel budget taken up by pensions increased from 27.6 percent in 1995 to 33.5 percent in 2002.

In addition, Brazilian universities employ large numbers of administrative and support staff whose remuneration swells the high cost of personnel. In this area, too, Brazil stands out as the Latin American university system with the highest proportion of nonteaching staff. In fact, Brazil is the only country with more administrative than teaching staff.

Unit costs hardly tell the full story. A thorough assessment of internal efficiency would first require hard analysis of the costs of producing individual graduates, as well as kinds of graduates and the economic output that graduates eventually contribute. Unfortunately, no recent studies have been done to determine, for example, the theoretical versus the actual time to complete

Figure E.4. Unit Costs in Selected Latin American Countries

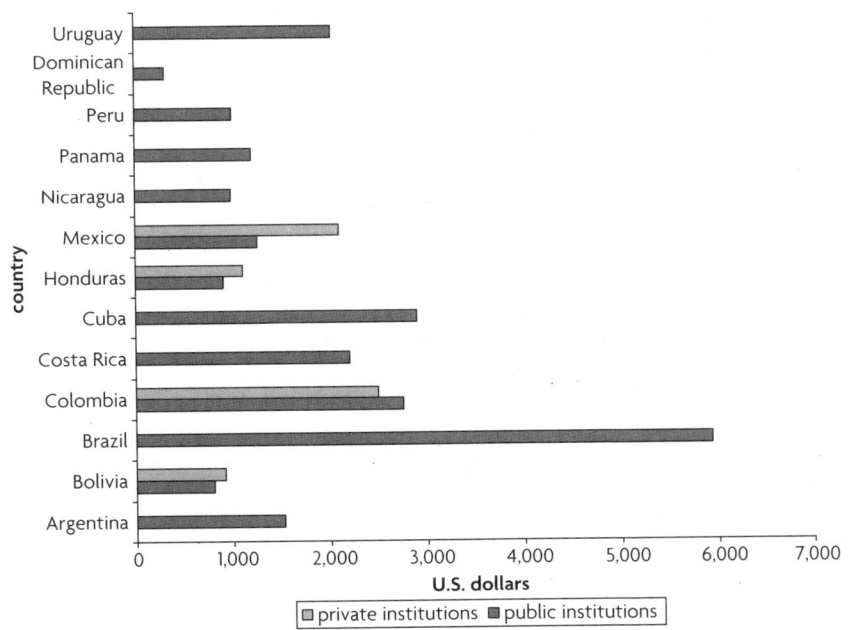

Source: IESALC 2006. Informe sobre la Educación Superior en América Latina y el Caribe 2000–05.

particular courses of study. Limited information available from UNICAMP reinforces the impression that considerable waste occurs despite the rigors of an exceptionally selective admission process. A survey of students admitted in 1994–97 revealed that only 72 percent had graduated by January 2005, another 26 percent had dropped out or been expelled, and the remaining 2 percent were still active (Pedrosa 2006).

One measurement of resource efficiency is to calculate a productivity ratio for each university, defined as the number of highly rated programs (levels six and seven in CAPES evaluations) divided by the number of professors with a PhD. The data show that the most productive universities in terms of research volume (USP and UNICAMP) are not the most efficient institutions when their teaching resources are factored in. Then UNIFESP and PUC-RIO are the most research efficient universities.

Below, table E.7 details the range of funding sources for Brazilian tertiary education institutions. Table E.8 lists several recent initiatives to promote research in national university systems around the world. While most of the initiatives are sponsored by OECD nations, the reader will note that Chile, China, and Korea also figure in the list.

Table E.7. Resource Diversification Matrix for Public Tertiary Education Institutions by Category and Source of Income

Category of income	Source of income				
	Government (national, state, municipal)	Students and families	Industry and services	Alumni and other philanthropists	International cooperation
Budgetary contribution					
General budget	X				
Dedicated taxes (lottery, tax on liquor sales, tax on contracts)	X				
Payroll tax			X		
Fees for instructional activities					
Tuition fees					
Degree/nondegree programs		X	X		
On-campus/distance-education programs		X	X		
Advance payments		X			
Charge-back	X				
Other fees (registration, labs, remote labs)		X			
Affiliation fees (colleges)			X		
Productive activities					
Sale of services			X		X
Consulting			X	X	X
Research			X		
Laboratory tests					
Patent royalties, share of spin-off profits, monetized patent royalties			X	X	
Operation of service enterprises (television, hotel, retirement homes, malls, parking, driving school, Internet provider, gym)			X		
			X		

(continued)

Table E.7. (*continued*)

Category of income	Government (national, state, municipal)	Students and families	Industry and services	Alumni and other philanthropists	International cooperation
Financial products (endowment funds, shares)			X		
Production of goods (agricultural and industrial)		X	X	X	
Thematic merchandise	X	X	X	X	X
Rental of facilities (land, classrooms, dormitories, laboratories, ballrooms, drive-throughs, concert halls, mortuary space)			X	X	
Sale of assets (land, residential housing)			X	X	
Fund raising					
Direct donations					
Monetary grants			X	X	X
Equipment			X	X	
Land and buildings	X			X	
Scholarships and student loans	X		X	X	X
Endowed chairs			X	X	
Indirect donations (credit card, percentage of gas sales, percentage of stock exchange trade, challenging grant)		X	X		
Tied donations (access to patents, share of spin-off profits)			X		
Concessions, franchising, licensing, sponsorships, partnerships (products sold on campus, names, concerts, museum showings, athletic events)			X		
Lotteries and auctions (scholarships)		X	X		
Loans					
Regular bank loans	X		X		
Bond issues		X	X	X	X

Source: Compiled by Jamil Salmi.

Table E.8. Recent Research "Excellence" Initiatives

Country	Number of target institutions and eligibility criteria	Resources allocated	Investment horizon
Germany Excellence Initiative 2006	40 graduate schools 30 clusters of excellence (universities and private sector) 10 top research universities	US$2.3 billion in total	Five-year funding Two rounds: 2006, 2007
Brain Korea 21 Program	Science and technology: 11 universities Humanities and social sciences: 11 universities Leading regional universities: 38 universities Professional graduate schools in 11 universities	US$1.17 billion in total	Seven years Two rounds in 1999
Korea Science and Engineering Foundation (KOSEF)	1) Science research centers (SRC)/engineering research centers (ERC): Up to 65 centers 2) Medical science and engineering research centers (MRC): 18 centers 3) National core research centers (NCRC): 6 centers funded in 2006	1) US$64.2 million per year 2) US$7 million per year 3) US$10.8 million per year	1) Up to nine years 2) Up to nine years 3) Up to seven years All three programs launched in FY2002 or FY2003
Japan Top-30 Program (Centers Of Excellence for 21st Century Plan)	31 higher-education institutions	US$150 million per year (Program total: ¥37.8 billion)	Five-year funding Launched in 2002 Three rounds: 2002, 2003, 2004
Japan Global Centers of Excellence Program	50–75 centers funded per year (five new fields of study each year)	¥50–¥500 million per center per year (~US$400,000 – US$4 million)	Five years Launched in 2007
European Commission, Framework Programme 7 (FP7)	TBD – determined by structure of research proposals (RFPs)	Based on number of RFPs with a "centre of excellence" structure The overall FP7 budget is ¥50.5 billion for 2007–13	Launched in 2007 2007–13
China 211 Project	100 higher-education institutions	US$18 billion in 7 years (US$400 million to funding world-class research departments)	Launched in 1996
China 985 Project	34 research universities	¥28.3 billion	1999–2001

(continued)

Table E.8. (*continued*)

Country	Number of target institutions and eligibility criteria	Resources allocated	Investment horizon
Chinese Academy of Sciences (CAS)		—	—
Institutes	Mathematics and physics: 15		
	Chemistry and chemical engineering: 12		
	Biological sciences: 20		
	Earth Sciences: 19		
	Technological sciences: 21		
	Others: 2		
Canada Networks of Centers of Excellence	23 currently funded networks of centers of excellence	Can$77.4 million per year since 1999	Operating since 1988
	16 previously funded networks	Can$47.3 million a year in 1997–99	Permanent program since 1997
		Can$437 million in total in 1988–98	
UK Funding for Excellent Units	Universities with the highest marks after the Research Assessment Exercise	US$8.63 billion disbursed after 2001 RAE	Five years for Research Council–funded centers
			Two rounds: 1996 and 2001 2008 RAE scheduled
Chile Millennium Science Initiative	Groups of researchers	Three science institutes: $1 million a year for 10 years	Every 5 years for nuclei and every 10 years for institutes
		5–12 science nuclei: US$250 thousand a year	
		US$25 million in total for 2000–2004	
Denmark (Globalization Fund)	Funds to be allocated competitively to research universities	US$1.9 billion between 2007 and 2012	Launched in 2006
NEPAD/Blair Commission for Africa (Proposed)	Revitalize Africa's institutions of higher education	US$500 million a year, over 10 years	Ten years
	Develop centres of excellence in science and technology, including African institutes of technology	Up to US$3 billion over 10 years	
Taiwan Development Plan for University Research Excellence	Selection and financial support of internationally leading fields	US$400 million	Four years

Source: Elaborated by Natalia Agapitova, Michael Ehst, and Jamil Salmi (last update March 9, 2007).

The Demographic Window of Opportunity

Although there are several labor market indicators of interest, the evolution of variables like population growth and labor supply are critical components of job creation and employment. An increasing population generally implies a larger labor supply that should be accompanied by sufficient labor demand or, in other words, more job creation. However, an increasing labor supply does not solely represent a challenge for the labor market. It is also a potential asset because increasing labor participation lowers society's dependency ratio and ensures labor revenues for pensions and other social expenditures—provided that this increasing labor force is employed and does not remain idle.

According to figure F.1, the proportion of Brazilians able to work (those older than 15 and younger than 64) has been increasing as a proportion of the total population, while Brazil's dependent population (those under 15 and above 64) has been consistently declining as a proportion of the labor force. Moreover, the proportion of the population able to work that actually participates in the labor force (active population) has risen from 66 percent to 74 percent during the past 25 years.[1] The population above 64 as a proportion of the labor force has remained relatively low and stable throughout the period at around 10 percent. In short, demographics are encouraging in terms of intergenerational transfers because there are more people able to work than in the past. However, this demographic "window of opportunity" will not last forever, because overall population growth is decreasing and the population above 64 is increasing.

Brazil is a populous country, with over 186 million people. However, population growth has declined from an annual 3 percent growth in the 1960s to 2 percent in the 1970s and 1980s before finally stabilizing at 1 percent in the 1990s. This is also true for the population growth of those aged 15–64. These demographics seem to predict lower pressures on the labor market in a 20-year scenario because labor supply is likely to decline as a consequence of lower population growth in the 1990s and early 2000s. In contrast, Brazil's labor market during the 1990s experienced the combined pressure of the late 1960s and 1970s baby boomers and higher female labor participation.

Figure F.1. Dependent Population and the Labor Force in Brazil, 1980–2005

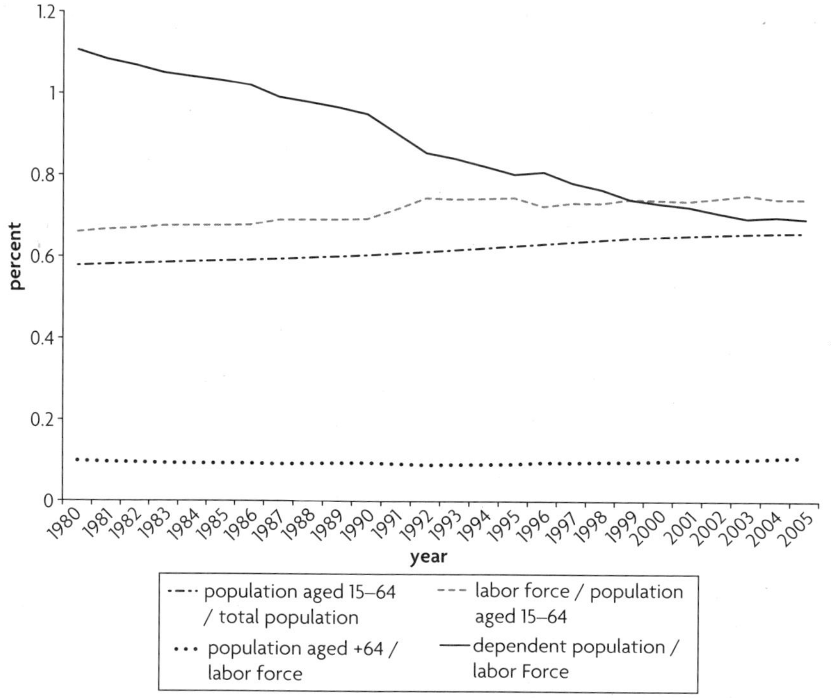

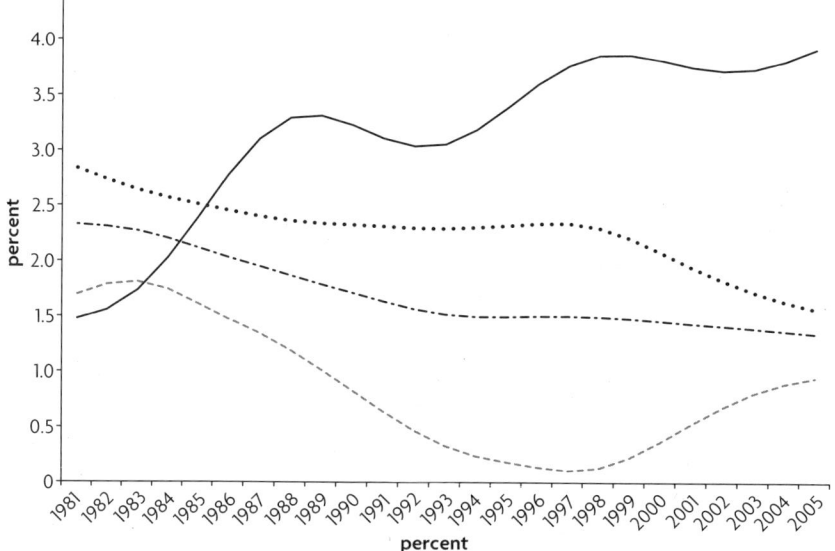

Source: Staff calculations with data from the WDI 2006.

Figure F.2. Population Trends by Age Group in Brazil, 1981–2005

Source: Staff calculations with data from the WDI 2006.

During the past 20 years, labor force growth has followed a more erratic and volatile path than population growth (figure F.3). There is a tendency to fall below 3 percent but with occasional jumps and, at other times, severe contractions. However, the population growth of the dependent population (those aged 0–14 and 65 and above) has also been declining. Female labor participation has been increasing in Brazil, and this is likely to continue in the foreseeable future. In 1980, women constituted only 31 percent of the labor force, but by 2005 they constituted 43 percent (*World Development Indicators* 2006).

Our analysis of labor supply contains two main features. At this moment, Brazil's labor force is much larger than it once was, so generating labor income to sustain social policies and old-age pensions is within reach, provided that there is sufficient employment creation. At the same time, low population growth will naturally decrease flows into the labor market. This will facilitate job creation, although female labor participation will moderate this effect. Brazil is undergoing what is commonly referred to as a process of demographic transition, a development that recently has been confirmed by Brazil's leading economic research institute (De Negri et al. 2006). In short, Brazil can take advantage of this demographic window for the next 20 years. After that, an increasing population in the over 64 age group and a shrinking labor force will put the system under stress, other things being equal.

Figure F.3. Population and Labor Supply in Brazil, 1981–2005

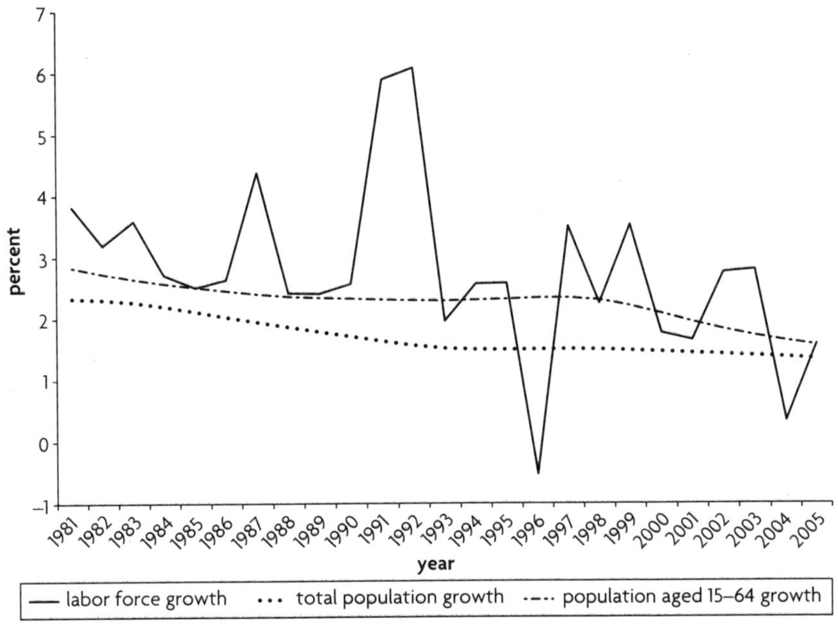

Source: Staff calculations with data from the WDI 2006.

Notes

Chapter 1

1. Brazil recently revised its national accounts from 1995 to 2006, showing an economy about 10 percent larger than previously estimated. Although the average real growth for 2003–06 rose to 4.1 percent (from 3.4 percent), average growth for 1996–2006 overall rose only to 2.5 percent (from 2.3 percent before the revision). All the estimates and calculations in this chapter include the newly revised national accounts data.

2. Knowledge-based industries include high- and medium-high-technology industries; communication services; finance, insurance, and other business services; and community, social, and personal services.

3. The technology intensity of trade presented in this table is based on the research and development (R&D) content of different sectors as derived from R&D spending in OECD countries and using input-output matrices to estimate the indirect R&D content of inputs.

4. This is supposed to characterize an economy at Brazil's level of GDP per capita and level of development. However, as can be seen by the rankings, Brazil is still struggling to improve the basic requirements that it should have mastered at an earlier stage.

5. For a good synopsis of the potential of services for growth in Latin American countries see Farrell and Remes (2007).

6. Knowledge-intensive services were defined as mail and telecommunications, finance and insurance, and business services (excluding real estate). The definition excludes government services, health, and education. A considerable part of health and education (which averaged 11 percent of GDP for OECD countries) could also be considered knowledge services, so these figures could be considered an underestimate of the share of knowledge services in the economy. See OECD (2005).

Chapter 2

1. See Rebelo 1990; and Barro 1991.
2. See, for example, Romer 1990; and Becker et al. 1990.
3. The data spanned more than 100 years for 23 of these countries.
4. Latin America is used here to refer to the countries included in the World Bank designation, LAC, which also includes Caribbean countries.
5. This growth decomposition exercise performed for Brazil highlights the importance (although it does not imply causality) of capital accumulation in the long run: growth of physical capital explains nearly half of GDP growth in all periods except the "lost decade" (1980s). During 1981–92, capital was used inefficiently, the result being negative rates of TFP change (−1.3 percent per year), which are mainly attributable to decreased capital productivity. Prior to that decade, capital contributed between 49 percent and 66 percent of GDP growth. Finally, in the "Real Era" (1993–2004), a much slower path of capital accumulation (3 percent per year, which yields a contribution to GDP growth of 1.52 percent) accounted for 52 percent of GDP growth.
6. Regarding the link between infrastructure and economic growth, Esfahani and Ramírez (2003) estimated a growth model for the average infrastructure (only in the power and telecommunications sectors) and GDP per capita growth rates of 75 countries for which complete data over any of the three decades 1965–75, 1975–85, and 1985–95 were available. Cross-country estimates showed that the contribution of infrastructure services to GDP is indeed substantial and generally exceeds the cost of such service provision. In addition, the steady-state elasticity of infrastructure with respect to total investment was found to be greater than one. Moreover, the widening infrastructure gap between East Asian and Latin American and Caribbean (LAC) countries accounted for nearly 25 percent of the GDP output gap between these two regions during 1980–2000 (Easterly 2000). Evidence exists that demonstrates not only the effects of infrastructure capital stock but also of infrastructure quality on the economic growth of LAC countries (Calderon and Serven 2005). In the Brazilian case, the infrastructure gap is estimated to be slightly smaller than the LAC average, although it is higher than in Argentina, Chile, and Uruguay (Calderon and Serven 2002). Data for Brazil confirms the positive relationship between infrastructure and economic growth in the long term, with telecommunications, electricity, and transportation as the most important sectors (Ferreira and Malliagros 1997). The same relationship was found for South Africa, where investment in infrastructure appears to have led economic growth during 1975–2001 (both directly and indirectly): while evidence of an output impact on infrastructure is weak, evidence of an infrastructure growth impact on output is robust (Fedderke et al. 2006).
7. In Brazil, 88 percent of commercial cases are appealed, compared with 30 percent in Mexico and 13 percent in Argentina.
8. Ruehl et al. 2005, provide an excellent summary of the analytical view on growth constraints in Brazil.

Chapter 4

1. The grouping of Brazil, Russia, India, and China was featured in a Goldman Sachs 2003 analysis, projecting the dominant position these economies may acquire in the global economy by 2050. See the Wikipedia BRIC entry for more detail.

2. See the Ministry of Science and Technology Web site link to *Indicadores de Ciência e Tecnologia* at www.mct.gov.br.

3. Ibid.

4. Patents granted in the United States in the past 17 years. The number of patents granted to Petrobrás by the U.S. Patent and Trademark Office between 1988 and 2005 is slightly lower than the total of 176 patents granted to it since 1976.

5. No updated numbers are available.

6. For more details, see Cruz and de Mello (2006).

7. See Beintema et al. 2001.

8. Lederman and Maloney (2003) estimated that the economic return on R&D in countries of Brazil's income level are high (around 65 percent), indicating that Brazil should be investing between two and eight times more in R&D than the 1990s levels.

9. This section draws heavily upon Alfred Watkins (2007).

10. Interestingly, if coincidentally, both Russia and Brazil fell nine places in WEF's overall competitiveness ranking between 2005/06 and 2006/07.

11. While analyzing productivity dispersion within sectors is a thought-provoking exercise, it must be noted that variance in productivity levels may be caused by factors such as economies of scale and intensive use of capital.

12. See World Bank (2007), "Unleashing India's Innovation Potential," for the productivity dispersion data on India and other countries.

13. The *Conselho Nacional de Desenvolvimento Científico e Technológico* (CNPq) was created in 1951, and two decades later the *Fundo Nacional de Desenvolvimento Científico e Tecnológico* (FNDCT) began operations. The former supported research mostly in the physical and natural sciences (directly through its institutes and indirectly through grants), while the latter, administered by FINEP (created in 1967), provided funding to boost graduate studies in universities in the 1980s and research activities in public enterprises (in addition to public research organizations) in the 1970s. The Ministry of Science and Technology was created as an overall coordinating body in 1985.

14. The original statute authorized the key incentive of an 8 percent corporate income tax write-off for R&D expenses, which the 1997 amended law reduced to 4 percent (including firms' expenditures with the *Programa de Alimentação do Trabalhador*). Although the incentives granted between 1994 and 2002 under Law 8.661/93 totaled R$1,158.2 million, corresponding to R$4,147.6 million of investments, only R$239.8 million were actually used during the period, while investments totaled R$3,338.6 million (all in current Brazilian reais). For an excellent discussion of the Brazilian R&D support system, particularly its different fiscal incentive regimes, see Confederação Nacional da Indústria 2005.

15. The MCT and the CCT were created by Decree No. 91,146/1985 and Law No. 9,257/1996, respectively.

16. The Coordination for the Improvement of Higher Education Staff (*Coordenação de Aperfeiçoamento de Pessoal de Nível Superior*, CAPES), attached to the Ministry of Education, is also responsible for improving the qualifications of university professors, mostly by financing postgraduate studies.

17. In addition to federal institutions, Brazil has several state-level institutions, such as the São Paulo State Institute for Technological Research (*Instituto de Pesquisas Tecnológicas do Estado de São Paulo*, IPT) and the São Paulo State Research Foundation (*Fundação de Amparo à Pesquisa do Estado de São Paulo*, FAPESP).

18. In 2005, the state of São Paulo invested about R$700 million in R&D. Figures for Rio de Janeiro, Minas Gerais, and Rio Grande do Sul were R$77 million, R$57 million, R$49 million, respectively.

19. For example, through Carta-Convite, FINEP publicly invites firms, as well as universities and research centers, to submit their project proposals. Funding is directed to public institutions and requires matching contributions by the private sector, which can also be financed through FINEP under its credit lines.

20. For an extensive examination of the performance of the funds since their creation, see Guimarães (2006).

Chapter 5

1. The Investment Climate Survey (ICS) is a comparative assessment undertaken annually by the World Bank and private partners. It uses a standard questionnaire to capture and quantify firms' real-world encounters with their national investment climates—the financial institutions, governance, business regulations, tax policies, labor relations, and technology that affect operations. Standardized data across a broad range of countries allows us to compare the "enabling-environments" of firms, both within Brazil and in other countries. A significant limitation of this database for our analysis is that the only questions it had on innovation in Brazil were whether firms had developed new products or updated product lines. It did not cover process innovation, which is the most prevalent type of innovation by firms in developing countries. However, this database is used because we have been able to undertake significant relevant analytical work on some critical relationships with it, whereas we have not had direct access to the PINTEC database.

2. See Viotti, Baessa, and Koeller (2005).

3. The PINTEC database analysis developed by Arbache (2005) combines several databases: (a) IBGE's Annual Industrial Survey (*Pesquisa Industrial Anual*, PIA) for firms' characteristics, (b) the Ministry of Labor and Employment's Annual Listing of Social Information (*Relação Anual de Informações Sociais*, RAIS) for variables related to the labor force, (c) IBGE's Industrial Survey–Technological Innovation (*Pesquisa Industrial–Inovação Tecnológica*, PINTEC) for information on innovation, (d) the central bank's Foreign Capital Census (*Censo do Capital Estrangeiro no Brasil*, CEB), and (e) the administrative database of the International Trade Secretariat (*Secretaria de Comércio Exterior*, Secex) for exporting information. For this report, the PINTEC database is more appropriate because it covers a much larger sample (over 72,000 firms), has a wider range of variables (including, in particular, a definition of innovation that covers both product and process innovations and many more questions about the information sources for innovation), and contains a time dimension.

4. See Romer (1990) and Aghion and Howitt (1997).

5. This finding could be explained by the sample's characteristics. Another possible explanation, suggested by the author, is that this coefficient would be capturing the effect of firms whose comparative advantage is in the production of goods intensive in unskilled labor and natural resources.

6. Independent variables included IC variables, plant control variables, and several dummy variables.

7. The drawback of this analysis is that its econometric specification is restricted to the IC variables, which are the sole explanatory variables.

8. The restricted case assumes that input-output elasticities are constant for all firms or are allowed to vary at the industry level (that is, the average cost share of each input is taken across the entire sample of plants from the seven countries). In the unrestricted case, the coefficients of the production function inputs are allowed

to vary industry by industry (that is, the cost share of each input is obtained for each of the nine manufacturing industries; and for each industry, plants were pooled from all the countries).

9. For detailed information on the econometric approach, see the technical annex of World Bank (2005a).

10. Note that certain reverse causality effects may be present. For example, it may be the case that more productive firms choose to provide external training.

11. This section is based on results of Correa et al. (forthcoming).

12. Other studies addressing these questions have been done for Chile, China, the Netherlands, and Sweden (Hall and Mairesse 2006).

13. In environments where capital markets tend to be imperfect, large firms tend to have greater possibilities to secure the resources needed for R&D activities. Phelps and Zoega (2001), working in a large sample of OECD countries, found that a well-developed stock market helps create profitable opportunities for entrepreneurs.

14. This finding is similar to that of Sbragia et al. (2004), who studied Brazilian firms using the database of the National Association of Research, Development, and Engineering in Innovative Firms (*Associação Nacional de P,D&E das Empresas Inovadoras*, ANPEI) during 1994–98.

15. For example, see Baldwin and Scott (1987) and Scherer and Ross (1990).

16. For more details on human capital, see chapter 6 of this report.

Chapter 7

1. A background discussion paper, videotape of speakers, and presentations from guests can be found at http://www.worldbank.org/stiglobalforum.

2. See, most recently, World Bank (2007a). Also see the Jobs Report (World Bank 2002a: Vol. 1), which recommended changes in labor regulations to achieve a more flexible and effective workforce, and see World Bank (2001) on pension reform.

3. See World Bank (2006b: Overview).

4. In past years, the Bank has assisted the Brazilian government to focus attention on some of these areas, including key studies to analyze the main challenges faced by the education sector. These include "A Call to Action, Combating School Failure in the Northeast of Brazil" (1997); "Brazil: Higher Education Sector Study" (2000); "Secondary Education in Brazil: Time to Move Forward" (2000); "Brazil: Teachers Development and Incentives" (2001); "Eradicating Child Labor in Brazil" (2001); "Brazil: Jobs Report" (2002); "Next Steps for Education in Four Selected States in Brazil" (2003); "An Assessment of the Bolsa Escola Programs" (2001); "Brazil: Early Child Development, A Focus on the Impact of Preschools" (2001); and "Brazil: Municipal Education, Resources, Incentives, and Results" (2002).

Appendix A

1. This section is based on Arbache (2005), which explored the relationship between innovation and exports, and the performance of manufacturing firms in Brazil. It combined the following databases: (a) IBGE's Annual Industrial Survey (*Pesquisa Industrial Anual*, PIA) for firms' characteristics, (b) the Ministry of Labor and Employment's Annual Listing of Social Information (*Relação Anual de Informações*

Sociais, Rais) for labor force variables, (c) IBGE's Industrial Survey–Technological Innovation (*Pesquisa Industrial–Inovação Tecnológica*, PINTEC) for information on innovation, (d) the central bank's Foreign Capital Census (*Censo do Capital Estrangeiro no Brasil*, CEB), and (e) the administrative database of the International Trade Secretariat (*Secretaria de Comércio Exterior*, Secex) for exporting information.

Appendix C

1. Sanguinetti (2005) found total employment to be a determinant of R&D expenditure per employee for Argentinean firms (a nonlinear relationship).

Appendix D

1. This ratio is computed by dividing total enrollment by full-time and part-time teachers, with the latter given a value of 0.5 (PISA Technical Manual 2003). This is not the same as class size, but it gives an idea of the overall teaching load.

Appendix E

1. For a critical assessment of league tables methodologies and policy usefulness, see Salmi and Saroyan (2007).

2. In 2000, 26 percent of Brazil's overall research population was employed in firms, compared with 70 percent in universities. By 2004, the disparity widened to 19 percent in firms and 77 percent in universities. By contrast, almost 70 percent of researchers in OECD countries are active in firms, and less than 25 percent in universities.

Appendix F

1. There is likely to be some upward bias in these percentages, since some of those in the labor force are below the age of 15 or above the age of 64.

References

Abrahão, J. 2005. "Financiamento e gasto público da educação básica no Brasil e comparacões com alguns países da OCDE e América Latina." *Educação & Sociedad* 26 (92): 841–58. Campinas: Brasil.

Adrogué, R., M. Cerisola, and G. Gelos. 2006. "Brazil's Long-Term Growth Performance: Trying to Explain the Puzzle." Working Paper No. 06/282. International Monetary Fund, Washington, DC.

Aghion, P., and P. Howitt. 1992. "A Model of Growth through Creative Destruction." *Econometrica* 60: 323–51.

Akinlo, A. E. 2005. "Impact of Macroeconomic Factors on Total Factor Productivity in Sub-Saharan African Countries." Working Paper No. 2005/39. World Institute for Development Economics Research, Helsinki, Finland.

Arbache, J. S. 2005. "Inovações tecnológicas e exportações afetam o tamanho e produtividade das firmas manufatureiras? Evidências para o Brasil." In *Inovações, padrões tecnológicos e desempenho das firmas industriais Brasileiras*. Ed. J. A. De Negri and M. S. Salerno. Brasília: IPEA.

Arbix, G. Forthcoming. "Innovative Firms in Three Emerging Economies: A Comparison between the Brazilian, Mexican, and Argentinean Industrial Elite."

Arraes, R. A., and V. K. Teles. 2003. "Differences in Long-Run Growth Path between Latin American and Developed Countries: Empirical Evidences." *Proceedings of the 31st Brazilian Economics Meeting*, No. C10.

Autor, D. H., F. Levy, and R. J. Murnane. 2003. "The Skill Content of Recent Technological Change: An Empirical Exploration." *Quarterly Journal of Economics* 118 (4): 1279–1333.

Bacha, L. E. 1977. "Issues and Evidence on Recent Brazilian Economic Growth." *World Development* 5 (1–2): 47–67.

Bacha, L. E., and R. Bonelli. 2004. "Accounting for Brazil's Growth Experience: 1940–2002." Discussion Paper No. 1018. IPEA, Brasília.

Bahia, L. D., and J. S. Arbache. 2005. "Diferenciacão salarial segundo criterios de desempenho das firmas industriais Brasilerias." In *Inovações, padrões tecnológicos e desempenho das firmas industriais Brasileiras*. Ed. J. A. De Negri and M. S. Salerno. Brasília: IPEA.

Baier, S. L., G. P. Dwyer, Jr., and R. Tamura. 2006. "How Important Are Capital and Total Factor Productivity for Economic Growth?" *Economic Inquiry* 44 (1): 23–49.

Baldwin, W. L., and J. T. Scott. 1987. *Market Structure and Technological Change.* Chur, Switzerland; New York: Harwood Academic Publishers.

Barbosa, N. H. 2001. "International Liquidity and Growth in Brazil." Working Paper No. 2001.04. Bernard Schwartz Center for Economic and Policy Analysis, New School University, New York.

Barro, R. J. 1991. "Economic Growth in a Cross-Section of Countries." *Quarterly Journal of Economics* 106 (2): 407–43.

Barro, R. 1996. "Determinants of Economic Growth: A Cross-Country Empirical Study." Working Paper No. 5698. National Bureau of Economic Research, Cambridge, Massachusetts.

Barros, R. P. de, R. Mendonça, D. Santos, and G. Quintaes. 2001. "Determinantes do Desempenho Educacional no Brasil." IPEA Working Paper No. 834. IPEA, Brasília.

Bartel, A. P. 2000. "Measuring the Employer's Return on Investments in Training: Evidence from the Literature." *Industrial Relations* 39 (3).

Becker, G. S., K. M. Murphy, and R. Tamura. 1990. "Human Capital, Fertility, and Economic Growth." *Journal of Political Economy* 98: S12–37.

Bedi, A. S., and J. H. Marshall. 2002. "Primary school attendance in Honduras." *Journal of Development Economics* 69: 129–53.

Beintema, N. M., A. F. D. Avila, and P. G. Pardey. 2001. "Agricultural R&D in Brazil: Policy, Investments, and Institutional Profile." IFPRI-Embrapa-Fontagro (International Food Policy Research Institute –Empresa Brasileira de Pesquisa Agropecuária–Fondo Regional de Tecnología Agropecuaria), Washington, DC.

Berg, J., C. Ernst, and P. Auer. 2006. *Meeting the Employment Challenge: Argentina, Brazil, and Mexico in the Global Economy.* London: Lynne Reinner.

Bertschek, I. 1995. "Product and Process Innovation as a Response to Increasing Imports and Foreign Direct Investment." *The Journal of Industrial Economics* XLIII (4): 341–57.

Biondi, R. L., and F. Felício. 2007. *Atributos escolares e o desempenho dos estudantes: uma análise em painel dos dados do Saeb.* Textos para Discussão No. 84. INEP (Instituto Nacional de Estudos e Pesquisas Educacionais Anísio Teixeira), Brasília.

Blundell, R., L. Dearden, C. Meghir, and B. Sianesi. 1999. "Human Capital Investment: The Returns from Education and Training to the Individual, the Firm, and the Economy." *Fiscal Studies* 20 (1).

Bon, G. C., and S. Gopinathan. 2006. "The Development of Education in Singapore since 1965." Background paper prepared for the East Asia Education Study Tour for African Policy Makers, June 18–30.

Bonelli, Regis. 1992. "Growth and Productivity in Brazilian Industries." *Journal of Development Economics* 39 (1): 85–109.

Calderon, C., and L. Serven. 2002. "The Output Cost of Latin America's Infrastructure Gap." Working Paper No. 186. Central Bank of Chile, Santiago.

———. 2005. "The Effects of Infrastructure Development on Growth and Income Distribution." World Bank and Inter-American Development Bank, Washington, DC.

Carnoy, M., A. Gove, S. Loeb, J. Marshall, and M. Socias. 2008. "How Schools and Students Respond to School Improvement Programs: The Case of Brazil's PDE." *Economics of Education Review* 27: 22–38.

Carnoy, M., A. Gove, and J. H. Marshall. 2007. *Cuba's Academic Advantage: Why Students Do Better in Cuban Schools.* Palo Alto: Stanford University Press.

Carnoy, M., and J. H. Marshall. 2005. "Comparing Cuban Academic Performance with the Rest of Latin America." *Comparative Education Review* 49 (2), 230–61.

Cohen, W. M., R. C. Levin, and D. C. Mowery. 1987. "Firm Size and R&D Intensity: A Reexamination." *Journal of Industrial Economics* 35: 543–65.

Confederação Nacional da Indústria. 2005. "Incentivos à inovação e à P&D no Brasil: Proposta de novo regime de apoio." Preliminary draft. Confederação Nacional da Indústria, Brasília.

Correa, P., I. S. Garcia, and H. Singh. Forthcoming. "Research, Innovation, and Productivity: Firm-Level Analysis for Brazil." World Bank report. World Bank, Washington DC.

Coulombe, S., J-F. Tremblay, and S. Marchard. 2004. "Literacy Scores, Human Capital, and Economic Growth in Fourteen OECD Countries." International Adult Literacy Survey Series. Statistics Canada and Human Resources Development Canada, Ottawa.

Crepon, B., E. Duguet, and J. Mairesse. 1998. "Research, Innovation and Productivity: An Econometric Analysis at the Firm Level." Working Paper No. 6696. National Bureau of Economic Research, Cambridge, Massachusetts.

CNI (National Confederation of Industry). 2005. "Incentivos à Inovação e à P&D no Brasil: Proposta de novo regime de apoio," Preliminary Draft. Brasília.

Cruz, C. H. B., and L. de Mello. 2006. "Boosting Innovation Performance in Brazil." Working Paper No. 532. Economics Department, OECD, Paris.

Dahlman, C., J. Routti, and P. Ylä-Anttila. 2005. *Finland as a Knowledge Economy: Elements of Success and Lessons Learned.* Overview. Washington, DC: Georgetown University Press and the World Bank. http://info.worldbank.org/tools/docs/library/201645/Finland_ES.pdf.

De Ferranti, D., and G. Perry. 2003. "Closing the Gap in Education and Technology." World Bank, Washington, DC.

De Gregorio, José. 1992. "Economic Growth in Latin America." *Journal of Development Economics* 39 (1): 59–84.

De Negri, F. 2006. "Determinantes da inovação e da capacidade de absorção nas firmas brasileiras: Qual a influência do perfil da mão-de-obra?" *Proceedings of the 34th Brazilian Economics Meeting,* Paper No. 100.

De Negri, J. A., F. de Negri, and D. Coelho, editors. 2006. *Tecnologia, exportação e emprego.* Brasília: IPEA.

De Negri, J. A., and M. S. Salerno, eds. 2005. *Inovações, padrões tecnológicos e desempenho das firmas industriais Brasileiras.* Brasília: IPEA.

Di Gropello, E., ed. 2006. *Meeting the Challenges of Secondary Education in Latin America and East Asia.* Washington, DC: The World Bank.

Dutta, S., and A. Lopez-Claros. 2005. *The Global Information Technology Report 2004–05: Efficiency in an Increasingly Connected World.* New York: Palgrave MacMillan.

Dutz, M. A., ed. 2007. *Unleashing India's Innovation: Toward Sustainable and Inclusive Growth.* Washington, DC: World Bank.

Easterly, W. 2000. "The Lost Decades: Explaining Developing Countries Stagnation 1980–98." Mimeographed document. World Bank, Washington, DC.

Easterly, W., and R. Levine. 2000. "It's Not Factor Accumulation: Stylized Facts and Growth Models." *IMF Seminar Series* 2000–12 (March): 1–52.

Elías, Víctor J. 1992. *Sources of Growth: A Study of Seven Latin American Economies.* Fundación del Tucumán and International Center for Economic Growth.

Escribano, A., and J. L. Guasch. 2004. "Assessing the Impact of the Investment Climate on Productivity Using Firm-Level Data: Methodology and the Cases of Guatemala, Honduras, and Nicaragua." Mimeographed document.

Escribano, A., N. Peltier-Thiberge, L. Garrido, and H. Singh. Forthcoming. "The Impact of Infrastructure on Competitiveness in Latin America: A Firm-Level Analysis Based on Investment Climate Assessments." Mimeographed document.

Esfahani, H. S., and M. T. Ramírez. 2003. "Institutions, Infrastructure, and Economic Growth." *Journal of Development Economics* 70 (2): 443–77.

FAPESP (Fundação de Amparo à Pesquisa do Estado de São Paulo). 2004. "Science, Technology, and Innovation in the State of São Paulo." Mimeographed document. FAPESP, São Paulo. http://www.fapesp.br.

Fajnzylber, P., and D. Lederman. 1999. "Economic Reforms and Total Factor Productivity Growth in Latin America and the Caribbean (1950–1995)." Policy Research Working Paper No. 2114. World Bank, Washington, DC.

Farrell, D., and J. Remes. 2007. "Tapping Latin America's Potential in Services." *McKinsey Quarterly* (May).

Fedderke, J. W., P. Perkins, and J. M. Luiz. 2006. "Infrastructural Investment in Long-Run Economic Growth: South Africa 1875–2001." *World Development* 34 (6): 1037–59.

Ferreira, P. C., and T. G. Malliagros. 1997. "Impactos productivos da infra-estrutura no Brasil: 1950–1995." Mimeographed document.

Ferreira, Pedro Cavalcanti, and J. L. Rossi. 2003. "New Evidence form Brazil on Trade Liberalization and Productivity Growth." *International Economic Review* 44 (4): 1383–405.

Ferreira, Pedro Cavalcanti, Samuel de Abreu Pessõa, and Fernando Veloso. 2006. "The Evolution of TFP in Latin America." Ensaios Econômicos No. 620. Escola de Pós Graduação en Economia da Fundação Getulio Vargas, Rio de Janeiro.

FIAS (Foreign Investment Advisory Service). 2001. "Brazil: Legal, Policy, and Administrative Barriers to Investment in Brazil—Volume I." FIAS-IFC/World Bank, Washington, DC.

Fuller, B., L. Dellagnelo, A. Strath, E.S.B. Bastos, M.H. Maia, A.L. Lopes de Matos, K.S. Portela, and S.L. Vieira. 1999. "How to Raise Children's Early Literacy? The Influence of Family, Teacher, and Classroom in Northeast Brazil." *Comparative Education Review* 43: 1–35.

Gibbons, M. 1998. "Higher Education Relevance in the 21st Century." Human Development Network, World Bank, Washington, DC.

Gomes, V., S. A. Pessôa, and F. A. Veloso. 2003. "Evolução da produtividade total dos fatores na economia Brasileira: Uma análise comparativa." *Pequisa e planejamento econômico* 33 (3): 389–434.

Gomes-Neto, J. B., and E. A. Hanushek. 1994. "Causes and Consequences of Grade Repetition: Evidence from Brazil." *Economic Development and Cultural Change* 43, 117–48.

Gopinathan, S. 1999. "Preparing for the Next Rung: Economic Restructuring and Educational Reform in Singapore." *Journal of Education and Work* 12 (3): 295–308.

Grilliches, Z. 1990. "Patent Statistics as Economic Indicators: A Survey." *Journal of Economic Literature* 28: 1661–707.

Guimarães, E. A. 2006. "Políticas de inovação: Financiamentos e incentivos." Discussion Paper No. 1212. IPEA, Brasília.

Hall, B. H., and J. Mairesse. 2006. "Empirical Studies of Innovation in the Knowledge-Driven Economy." Working Paper No. 12320. National Bureau of Economic Research, Cambridge, Massachusetts.

Hammergren, L. 2004. "Brazil: Making Justice Count: Measuring and Improving Judicial Performance in Brazil." World Bank Report No. 32789-BR. World Bank, Washington, DC.

Hanushek, E., and V. Lavy. 1994. "School Quality, Achievement Bias and Dropout Behaviour in Egypt." Living Standards Measurement Survey Study Working Paper 107. World Bank, Washington, DC.

Hanushek, E. A., and L. Wößmann. 2007. "The Role of Education Quality in Economic Growth." Policy Research Working Paper No. 4122. World Bank, Washington, DC.

Helpman, E. 2004. *The Mystery of Economic Growth*. Cambridge, MA: Harvard University Press.

IBGE (Instituto Brasileiro de Geografia e Estadística). 2004. PNAD (Pesquisa Nacional por Amostra de Domicílios). IBGE, Rio de Janeiro.

IDB (Inter-American Development Bank). 2006. "Education, Science, and Technology in Latin America and the Caribbean: A Statistical Compendium of Indicators." IDB, Washington, DC. http://www.iadb.org/sds/doc/EducationScienceandTechnology.pdf.

IESALC (International Institute for Higher Education in Latin America and the Caribbean). 2006. *Informe sobre la educación Superior en América Latina y el Caribe 2000–2005: La metamorfosis de la educación superior*. UNESCO, Caracas.

Ioschpe, G. 2004. *A Ignorância Custa um Mundo*. São Paulo: Editora Francis.

Ioschpe, G. 2006. "Custos e Determinantes de Repetência e Evasão Escolares no Brasil." Unpublished policy note commissioned by the World Bank. World Bank, Washington, DC.

Ioschpe, G., and C. de Moura. 2007. "La remuneración de los maestros en America Latina: ¿Es baja? ¿Afecta la Enseñanza?" PREAL, Washington, DC.

Jaffe, A. B., and M. Trajtenberg. 2002. *Patents, Citations, and Innovations: A Window on the Knowledge Economy*. Cambridge, MA: MIT Press.

Janz, N., H. Loof, and B. Peters. 2003. "Firm-Level Innovation and Productivity: Is There a Common Story across Countries?" Working Paper No. 24. Center for European Economic Research, Mannheim.

JBIC (Japan Bank for International Cooperation). 2005. Sector Study for Education in Brazil. JBIC, Tokyo.

Jones, C. I. 2002. "Sources of U.S. Economic Growth in a World of Ideas." *American Economic Review* 92: 220–39.

Keller, W. 2002. "Geographic Localization of International Technology Diffusion." *American Economic Review* 92: 120–42.

Koeller, P., and A. R. Baesa. 2005. "Inovação tecnológica na industria Brasileira." In *Inovações, padrões tecnológicos e desempenho das firmas industriais Brasileiras*. Ed. J. A. De Negri and M. S. Salerno. Brasília: IPEA.

Krugman, P. 1994. "The Myth of Asia's Miracle." *Foreign Affairs* (November/December).

Lederman, D., and W. Maloney. 2003. "R&D and Development." Mimeographed document. Office of the Chief Economist for LAC, World Bank, Washington, DC.

Lederman, D., and L. Saenz. 2004. "Innovation and Development around the World." Policy Research Working Paper No. 3774. World Bank, Washington, DC.

Levine, R., and D. Renelt. 1992. "A Sensitivity Analysis of Cross-Country Growth Regressions." *American Economic Review* 82 (4).

Levy, F., and R. J. Murnane. 2004. *The New Division of Labor: How Computers are Creating the Next Job Market.* Princeton, NJ: Princeton University Press.

Lloyd, C. B., S. El Tawila, W. H. Clark, and B. S. Mensch. 2003. "The Impact of Educational Quality on School Exit in Egypt." *Comparative Education Review* 47: 444–67.

Loayza, N., P. Fajnzylber, and C. Calderón. 2004. "Economic Growth in Latin America and the Caribbean: Stylized Facts, Explanations, and Forecasts." Working Paper No. 65. Central Bank of Chile, Santiago.

Lofts, C., and J. Loundes. 2000. "Foreign Ownership, Foreign Competition and Innovation in Australian Enterprises." Working Paper No. 20/00. Melbourne Institute of Applied Economic and Social Research, The University of Melbourne, Melbourne.

Love, J. H., B. Ashcroft, and S. Dunlop. 1996. "Corporate Structure, Ownership, and the Likelihood of Innovation." *Applied Economics* 28: 737–46.

Marshall, J. H. 2003. "If You Build it Will They Come? The Effects of School Quality on Primary School Attendance in Rural Guatemala." Doctoral dissertation. Stanford University School of Education, Stanford, CA.

Marshall, J. H. 2003b. "Grade Repetition in Honduran Primary Schools." *International Journal of Educational Development* 23(6), 591–605.

Marshall, J. H. Forthcoming. "School Quality and Learning Gains in Rural Guatemala." *Economics of Education Review.*

Meyer, J., F. Ramirez, and E. Schofer. 2000. "The Effects of Science on National Economic Development, 1970–1990." *American Sociological Review* 65: 877–98.

Ministry of Education, People's Republic of China. 2005. *Education in China.* Beijing: Ministry of Education.

Mohnen, P., and M. Dagenais. 2002. "Towards an Innovation Intensity Index: The Case of CIS-I in Denmark and Ireland." In *Innovation and Firm Performance: Econometric Explorations of Survey Data*, pp. 3–30. Ed. A. Kleinknecht and P. Mohnen. New York, NY: Palgrave.

Moreira, M. M. 2004. "Brazil's Trade Liberalization and Growth: Has It Failed?" Occasional Paper No. 24. IDB-INTAL (Inter-American Development Bank-Institute for the Integration of Latin America and the Caribbean), Washington, DC.

OECD (Organisation for Economic Co-Operation and Development). 2000. *Science, Technology and Industry Outlook.* Paris: OECD.

———. 2001. *Using Knowledge for Development: The Brazilian Experience.* Paris: OECD.

———. 2005. *OECD Science, Technology and Industry: Scoreboard 2005.* Paris: OECD.

Pedrosa, R. H. L. 2006. "Educational and Socioeconomic Background of Undergraduates and Academic Performance: Consequences for Affirmative Action Programs at a Brazilian Research University." Presentation at the IMHE/OECD General Conference on Values and Ethics in Higher Education, Paris, September, 11–13, 2006.

Phelps, E., and G. Zoega. 2001. "Structural Booms: Productivity Expectations and Asset Valuations." *Economic Policy* 32 (Spring).

Pinheiro, A. C., I. Gill, L. Servén, and M. Thomas. 2004. "Brazilian Economic Growth, 1900–2000: Lessons and Policy Implications." Inter-American Development Bank, Washington, DC.

Porter, M. E. 1990. *The Competitive Advantage of Nations.* New York: Free Press.

Pritchett, L. 1996. "Where Has All the Education Gone?" Policy Research Working Paper No. 1581. World Bank, Washington, DC.

Ranieri, N. B. S. 2006. "Aspectos Jurídicos da Autonomia Universitária no Brasil." Instituto de Estudos Avançados da Universidade de São Paulo, São Paulo. http://www.iea.usp.br/observatorios/educacao.

Rebelo, S. 1990. "Long-Run Policy Analysis and Long-Run Growth." Working Paper No. 3325. National Bureau of Economic Research, Cambridge, Massachusetts.

Rodríguez, A., and C. A. Herrán. 2000. "Secondary Education in Brazil: Time to Move Forward." World Bank and Inter-American Development Bank, Washington, DC.

Romer, P. M. 1990. "Endogenous Technological Change." *Journal of Political Economy* 98 (5): S71–102.

Ruehl, C., M. Thomas, J. Revilla, and A. Vivanco. 2005. "Growth Diagnostics: Is Saving the Main Binding Constraint for Brazilian Growth?"

Sá, C. 2005. "Research Policy in Emerging Economies: Brazil's Sector Funds." *Minerva* 43: 245–63.

Salmi, J., and A. Saroyan. 2007. "League Tables as Policy Instruments: Uses and Misuses." *Higher Education Management and Policy* 19 (2).

Salomon, R., and J. M. Shaver. 2005. "Learning by Exporting: New Insights from Examining Firm Innovation." *Journal of Economics and Management Strategy,* 14 (2): 431–60.

Sanguinetti, P. 2005. "Innovation and R&D Expenditures in Argentina: Evidence from a Firm-Level Survey." Universidad Torcuato Di Tella, Department of Economics, Buenas Aires. http://www.crei.cat/activities/sc_conferences/23/papers/sanguinetti.pdf.

Sbragia, R., N. Menezes-Filho, and J. Jensen. 2004. "Os determinantes dos gastos em P&D no Brasil: Uma análise com dados em painel." *Estudos Econômicos* (IPE/USP) 34 (4): 661–91.

Scherer, F. M., and D. Ross. 1990. *Industrial Market Structure and Economic Performance,* 3rd ed. Boston, MA: Houghton Mifflin.

Schwartzman, S., and C.M. Castro. 2005. *Reforma da Educação Superior: uma visão crítica.* Brasília: Funadesp.

Schumpeter, J. A. 1942. *Capitalism, Socialism, and Democracy.* New York and London: Harper and Brothers.

Singh, N., and H. Trieu. 1996. "The Role of R&D in Explaining Total Factor Productivity Growth in Japan, South Korea and Taiwan." Working Paper #362. Department of Economics, University of California, Santa Cruz, CA.

Siqueira, E. 2006. "Vinte obstáculos ao crescimento sustentável." *O Estado de S. Paulo* 24 (December).

Solow, R. 1956. "A Contribution to the Theory of Economic Growth." *Quarterly Journal of Economics* 50: 65–94.

Solow, R. M., 2001. "What Have We Learned from a Decade of Empirical Research on Growth? Applying Growth Theory across Countries." *World Bank Economic Review* 15 (2): 283–88.

Souitaris, V. 2002. "Firm-Specific Competencies Determining Technological Innovation: A Survey in Greece." *R&D Management* 32: 61–77.

Sowell, T. 2004. *Affirmative Action Around the World: An Empirical Study.* New Haven, CT: Yale University Press.

Stein, M. K., M. S. Smith, M. A. Henningsen, and E. A. Silver. 2000. *Implementing Standards-Based Mathematics Instruction: A Casebook for Professional Development.* New York: Teachers College Press.

Tafner, Paulo, ed. 2006. "O Estado de uma Nação." IPEA, Rio de Janeiro, Brasil.

Tendler, J. 2002. "The Fear of Education." Paper presented at the 50th anniversary of the Bank of the Northeast and at BNDES, Rio de Janeiro, July.

Tsang, M. 1996. "The Financial Reform of Basic Education in China." *Economics of Education Review* 15 (4): 423–44.

UNCTAD (United Nations Conference on Trade and Development). 2005. *World Investment Report 2005—Transnational Corporations and the Internationalization of R&D.* Geneva and New York: United Nations. http://wits.worldbank.org/witsweb.

Viotti, E. B., A. R. Baessa, and P. Koeller. 2005. "Perfil da inovação na indústria Brasileira: Uma comparação internacional." In *Inovações, padrões tecnológicos e desempenho das firmas industrias Brasileiras*. Ed. J. A. De Negri and M. S. Salerno. Brasília: IPEA.

Watkins, Alfred. 2007. "From Knowledge to Wealth: Transforming Russian Science and Technology for a Modern Knowledge Economy." Policy Research Working Paper No. 2974. World Bank, Washington, DC.

World Bank. 1999. *World Development Report 1998/1999: Knowledge for Development*. Washington, DC: World Bank.

———. 2000. "Brazil: Higher Education Sector Study." Report No. 19392. Human Development Department, Latin America and the Caribbean Region, World Bank, Washington, DC.

———. 2001. "Brazil: Critical Issues in Social Security." World Bank, Washington, DC.

———. 2002a. "Brazil—Jobs Report." Report No. 24408. Vols. 1&2. World Bank, Washington, DC.

———. 2002b. "Brazil: The New Growth Agenda." Report No. 22950. World Bank, Washington, DC.

———. 2004a. "Azerbaijan: Education Concept Policy Note." Report No. P102283. World Bank, Washington, DC.

———. 2004b. "Brazil: Access to Financial Services." Report No. 27773. World Bank, Washington, DC.

———. 2004c. "Brazil: First Programmatic Loan for Sustainable and Equitable Growth." Report No. 27507. World Bank, Washington, DC.

———. 2004d. "Brazil: Making Justice Count. Measuring and Improving Judicial Performance in Brazil." Report No. 32789. World Bank, Washington, DC.

———. 2004e. "Brazil: Trade Policies to Improve Efficiency, Increase Growth and Reduce Poverty." Report No. 24285. World Bank, Washington, DC.

———. 2005a. *Brazil: Investment Climate Assessment*, Vols. I & II. Washington, DC: World Bank.

———. 2005b. "Brazil: Second Programmatic Loan for Sustainable and Equitable Growth." Report No. 36059. World Bank, Washington, DC.

———. 2006a. "Brazil: Interest rates and Intermediation Spreads." Report No. 36628. World Bank, Washington, DC.

———. 2006. Development Data Platform.

———. 2006b. "Doing Business in Brazil." Report No. 36881. World Bank, Washington, DC.

———. 2006c. "How to Revitalize Infrastructure Investments in Brazil: Public Policies for Better Private Participation." Report No. 36624. World Bank, Washington, DC.

———. 2006d. *World Development Indicators*. Washington, DC: World Bank.

———. 2007a. "Brazil: Improving Fiscal Circumstances for Growth." Report No. 36595. Vols. 1&2. World Bank, Washington, DC.

———. 2007b. "Education, Training, and Innovation: The Empirical Link in the Brazilian Case." Mimeographed document. World Bank, Washington, DC.

World Bank Institute. 2006. "Does Training Work? Re-Examining Donor-Sponsored Training Programs in Developing Countries." *Capacity Development Briefs*, February (15).

World Development Indicators. Various years. Washington, DC: World Bank.

World Economic Forum. 2006. *The World Economic Forum Global Competitiveness Report*. www.wefourm.org.

World Bank and UNCTAD, World Integrated Trade Solution (WITS) database. http://wits.worldbank.org/witsweb/.

Index

Boxes, figures, and tables are indicated by b, f, and t, respectively.

ECO-AUDIT
Environmental Benefits Statement

The World Bank is committed to preserving endangered forests and natural resources. The Office of the Publisher has chosen to print **Knowledge and Innovation for Competitiveness in Brazil** on recycled paper with 30 percent post-consumer waste, in accordance with the recommended standards for paper usage set by the Green Press Initiative, a nonprofit program supporting publishers in using fiber that is not sourced from endangered forests. For more information, visit www.greenpressinitiative.org.

Saved:
- 9 trees
- 6 million BTUs of total energy
- 765 pounds of net greenhouse gases
- 3,176 gallons of waste water
- 408 lbs of solid waste